COOL THERMODYNAMICS

THE ENGINEERING AND PHYSICS OF
PREDICTIVE, DIAGNOSTIC AND OPTIMIZATION METHODS FOR COOLING
SYSTEMS

JEFFREY M GORDON
Ben-Gurion University of the Negev, Israel

KIM CHOON NG
National University of Singapore

Viva Books

New Delhi | Mumbai | Chennai | Kolkata | Bangalore | Hyderabad | Kochi

First Indian Edition 2008

VIVA BOOKS PRIVATE LIMITED

- 4737/23 Ansari Road, Daryaganj, New Delhi 110 002
  E-mail: vivadelhi@vivagroupindia.net, Tel. 42242200
- Plot No. 76, Service Industries, Shirvane, Sector 1, Nerul, Navi Mumbai 400 706
  E-mail: vivamumbai@vivagroupindia.net, Tel. 27721273, 27721274
- Jamals Fazal Chambers, 26 Greams Road, Chennai 600 006
  E-mail: vivachennai@vivagroupindia.net, Tel. 28290304, 28294241
- B-103, Jindal Towers, 21/1A/3 Darga Road, Kolkata 700 017
  E-mail: vivakolkata@vivagroupindia.net, Tel. 22836381, 22816713
- 7, Sovereign Park Apartments, 56-58, K. R. Road, Basavanagudi, Bangalore 560 004
  E-mail: vivabangalore@vivagroupindia.net, Tel. 26607409, 26607410
- 101-102 Mughal Marc Apartments, 3-4-637 to 641, Narayanguda, Hyderabad 500 029
  E-mail: vivahyderabad@vivagroupindia.net, Tel. 27564481, 27564482
- First Floor, Beevi Towers, SRM Road, Kaloor, Kochi 682 018, Kerala
  E-mail: vivakochi@vivagroupindia.net, Tel: 0484-2403055, 2403056

www.vivagroupindia.com

Published by arrangement with

Cambridge International Science Publishing
7, Meadow Walk, Great Abington
Cambridge CB1 6AZ, UK

ISBN-10: 81-309-0951-0
ISBN-13: 978-81-309-0951-6

Published by Vinod Vasishtha for Viva Books Private Limited, 4737/23, Ansari Road, Daryaganj, New Delhi-110 002. Printed and bound by Anand Sons, Delhi - 110 092.

Contents

About the Authors

Jeffrey M. Gordon

Prof. Gordon was born in 1949 in the USA. Currently, he holds the rank of professor at Ben-Gurion University of the Negev (Israel), in the Department of Energy & Environmental Physics, Sede Boqer Campus, and the Department of Mechanical Engineering, Beersheva Campus. He received his Ph.D. from Brown University in 1976.

Prof. Gordon has authored over 120 papers in international peer-reviewed journals in the areas of: the engineering and physics of cooling systems, finite-time thermodynamics, nonimaging optics, biomedical optics, and solar energy. He is an Associate Editor of the journal Solar Energy; editor of the International Solar Energy Society Background Paper Series, and former associate editor of the ASME J. of Solar Energy Eng., Progress in Photovoltaics and Advances in Thermodynamics.

Prof. Gordon is also a member of the board of reviewers of over a dozen additional leading journals, including Journal of Applied Physics, American Journal of Physics, International Journal of Heat and Mass Transfer, Applied Optics, Solar Energy Materials, and Journal of the Optical Society of America.

Kim Choon Ng

Prof. Ng was born in 1952 in Malaysia. He is now an Associate Professor, National University of Singapore, Department of Mechanical & Production Engineering, Singapore. He received his Ph.D. in 1980, from the University of Strathclyde, United Kingdom.

Prof. Ng has published over 40 papers in international peer-reviewed journals in the areas of: solar energy, chiller modeling and experimental testing, and two-phase flow.

He is a member of the board of reviewers of the International Journal of Refrigeration, Solar Energy, Heat Transfer Engineering and Applied Thermal Engineering, a Chartered Engineer (UK) and a registered Professional Engineer (PEng) in Singapore.

In the specific area of this book, Gordon and Ng have co-authored 13 papers in the following journals during the years 1994-2000: Journal of Applied Physics, International Journal of Refrigeration, Applied Thermal Engineering, International Journal of Heat and Mass Transfer and Solar Energy. Separately, they have authored more than two dozen additional articles on cooling systems, heat engines and chemical converters in leading journals.

Preface

Being familiar, but not *too* familiar, with a discipline can have its benefits. The complexities of cooling systems can be intimidating to anyone intent upon trying to develop relatively simple, analytic modeling procedures that offer diagnostic, predictive and optimization capabilities. In fact, an intimate familiarity with even the most common cooling devices such as building air conditioners and household refrigerators can dissuade even the ambitious researcher or practitioner from such tasks. This may partly explain why the analysis and modeling of cooling and refrigeration systems have been tackled with massive simulation techniques or largely empirical methods that forego the hope of capturing the essential physics of the problem in succinct terms.

Because we were not fully versed in every intricacy of these problems, we naively embarked upon the mission of developing uncomplicated models and procedures that could succeed in several of the key aims currently satisfied only with the nominally extreme approaches noted above. With some basic, unsophisticated engineering and physics, we found that surprisingly accurate and powerful tools emerged. Most of these results have been published in the journals during the past 5 years. At the encouragement of colleagues and cooling engineers, we felt it worthwhile to collate the lessons learned, the models derived, the experimental case studies, and the perspective of several years' experience with these results in book form. These recent advances are sandwiched between introductory material on chiller fundamentals and closing thoughts about challenges for future work.

The manner in which the book could be used in industrial workshops, university courses and other instructional settings, and the audiences to whom this book is tailored, are elaborated upon in Section D2 of Chapter 1. Toward guiding readers through much of the background material on cooling systems, and toward enabling them to gain a firm grasp on the recent progress from the journal papers, we have included more than a dozen tutorial examples. The tutorials are intended to assist the reader in translating the concepts and equations into readily-implemented design and diagnostic tools.

The research upon which much of this book is based evolved from a rewarding and gratifying collaboration between us, that started during the sabbatical year that one of us (JMG) spent with the other (KCN) at the National University of Singapore. It is a pleasure to acknowledge our partner in several of those research efforts, Hui Tong Chua, who was working on his master's and PhD degrees during the period of those research programs.

JMG also expresses his appreciation to his family, but not simply for the usual reasons of patient support. To my daughters (Shere, Nirit and Rona) and my wife (Yocheved) go my gratitude for always forcing me to try to explain engineering and scientific notions in lay terms that can be comprehended without a formal scientific education. Those challenges enabled me to develop a more profound understanding of the material presented in this book. I thank them for their encouragement and forbearance during this undertaking, and dedicate this book to them.

KCN would also like to dedicate this book to his wife (Linda) and children (Suzanne, Joseph and Sophia). Their unwavering love, devotion and support have made its completion possible.

Both of us (JMG and KCN) hope that this book will serve not only as a guide and educational tool for practicing engineers, university students and researchers, but will also serve as a first step in the direction of a universal thermodynamic modeling approach for cooling devices of all sorts: for elucidating their thermodynamic behavior, offering practical diagnostic tools, and providing optimization tools with which future generations of cooling systems can be designed and improved.

NOMENCLATURE

A	overall heat exchanger heat transfer area
A_i, A_o	tube inner/outer surface area
A_j	constants characterizing a reciprocating chiller in the quasi-empirical model (j = 1-3)
AHE	internal absorber heat exchanger
B_j	constants characterizing an absorption chiller in the quasi-empirical model (j = 1,2)
C	specific heat
COP	coefficient of performance (ratio of useful effect output to power input)
CR	circulation flow rate ratio (ratio of solution mass flow rate at the absorber to refrigerant mass flow rate)
D	tube diameter
E	heat exchanger effectiveness
E	internal energy
GAX	internal generator-absorber heat exchanger
GHE	internal generator heat exchanger
h	specific enthalpy
h_t	heat transfer coefficient
h_X	heat exchanger contribution in expression for 1/COP
H	enthalpy
I	electrical current
IPLV	integrated part load value (method for estimating long term performance of centrifugal chillers)
K	thermal conductance
L	tube length
LMTD	log-mean temperature difference in a heat exchanger
M_j	shorthand notation for mCE product in heat exchanger j
m	mass flow rate
$(mCE)'$	heat exchanger thermal inventory per unit of refrigerant charge
n	number of tubes
p	pressure
P_{in}	input power
PAT	process average temperature
Q	heat transfer

Q'	heat transfer in the internal perspective of the vortex tube chiller
Q_{cold}	heat removal to cold reservoir
Q_{hot}	heat input from hot reservoir
Q_{input}	total heat input
Q_{eqv}^{leak}	equivalent heat leak parameter of a chiller
Q_j^{leak}	heat leak at component j
Q_{net}^{leak}	net (weighted) heat leak
Q_{reject}	total heat rejection
q_j	additional heat losses that stem from internal losses in component j
R	effective thermal resistance of heat exchangers
R_j	thermal resistance of heat exchanger j
R_{el}	electrical resistance
Rton	refrigeration ton
s	specific entropy
S	entropy
S_{int}^k	rate of internal entropy production for component k
T	temperature
T_j^{in}	coolant inlet temperature at component j
T_j^{out}	coolant outlet temperature at component j
T_{av}^s	entropic average temperature
T_{abs}	refrigerant temperature at the absorber
$T_{adsorber}$	refrigerant temperature at the adsorber
T_c	temperature of the cold air extracted from the vortex-tube chiller
T_{cold}	temperature of cold reservoir
T'_{cold}	refrigerant temperature at the cold reservoir
T_{cond}	refrigerant temperature at the condenser
T_{dep}	refrigerant temperature at the dephlegmator
$T_{desorber}$	refrigerant temperature at the desorber
T_{evap}	refrigerant temperature at the evaporator
T_{gen}	refrigerant temperature at the generator
T_h	temperature of hot air extracted from the vortex-tube chiller
T_{hot}	temperature of hot reservoir
T'_{hot}	refrigerant temperature at the hot reservoir
T_o	temperature of the cold gas after expansion in the vortex-tube chiller
T_{pl}	temperature of the plenum (entrance) air in the vortex-tube chiller

U	overall heat exchanger heat transfer coefficient
V	volumetric flow rate
v	specific volume
W	work input
W'	work input in the internal perspective in the vortex-tube chiller
X	mass fraction in solution
Y	mass fraction in the vapor
y	cold fraction in vortex-tube chiller
α	differential thermoelectric power coefficient
ΔE	change in internal energy over one cycle
ΔS	change in entropy over one cycle
ΔS_{int}	rate of internal entropy production
ΔS_{leak}	rate of entropy production due to heat leak
δS_j	rate of entropy production per refrigerant charge for component j in the relative residence time analysis
δJ	experimental uncertainty in generalized variable J ($J = Q_{evap}$, P_{in}, m, ΔT)
ΔS_u	entropy production in the universe (chiller plus reservoirs)
ΔT	temperature change in a given process
ε_j	entropy terms for testing predictions of endoreversible chiller models (j = 1-4)
ρ	density
ξ	fraction of total heat rejection effected at the condenser in an absorption chiller
κ	a constant characterizing a heat exchanger (for coolant flow-rate dependence)
μ	chemical potential
Ξ_j	relative residence time (also relative refrigerant charge) for refrigerant in component j
ψ	fraction of total heat input accepted at the generator in an absorption heat transformer

CONVERSION TABLE

power (energy rate of change)

$1\ kW = 1000\ W = 3412\ Btu\ h^{-1}$ — $1\ Btu\ h^{-1} = 0.0002931\ kW$

$1\ kW = 0.2844\ Rton$ — $1\ Rton = 3.517\ kW$

COP (dimensionless)

$$COP = \frac{3.517}{\frac{kW}{Rton}} \qquad \frac{kW}{Rton} = \frac{3.517}{COP}$$

thermal conductance or entropy rate of change

$1\ kW\ K^{-1} = 1895\ Btu\ h^{-1}\ °F^{-1}$ — $1\ Btu\ h^{-1}\ °F^{-1} = 0.000528\ kW\ K^{-1}$

specific enthalpy

$1\ kJ\ kg^{-1} = 0.430\ Btu\ lbm^{-1}$ — $1\ Btu\ lbm^{-1} = 2.33\ kJ\ kg^{-1}$

specific heat or specific entropy

$1\ kJ\ kg^{-1}\ K^{-1} = 0.239\ Btu\ lbm^{-1}\ °F^{-1}$ — $1\ Btu\ lbm^{-1}\ °F^{-1} = 4.19\ kJ\ kg^{-1}\ K^{-1}$

temperature

$T(K) = T(°C) + 273.15$ — $T(°F) = 32 + 1.8\ T(°C)$

$$T(°C) = \frac{5\{T(°F) - 32\}}{9}$$

$T(R) = 459.67 + T(°F)$

pressure

$1\ kPa = 0.01\ bar = 0.009869\ atm = 20.886\ lbf\ ft^{-2}$ — $1\ lbf\ ft^{-2} = 0.04788\ kPa$

volumetric flow rate

$1\ l\ s^{-1} = 0.001\ m^3\ s^{-1} = 2.119\ ft^3\ min^{-1}$ (cfm) — $1\ cfm = 0.4719\ l\ s^{-1}$

$1\ l\ s^{-1} = 15.85\ gpm$ — $1\ gpm = 0.0631\ l\ s^{-1}$

mass flow rate

$1\ kg\ s^{-1} = 2.205\ lbm\ s^{-1}$ — $1\ lbm\ s^{-1} = 0.454\ kg\ s^{-1}$

Chapter 1

WHAT THE BOOK HAS TO OFFER AND THE INTENDED AUDIENCES: MODELING, DIAGNOSING AND OPTIMIZING COOLING DEVICES

"Whatever you do will be insignificant, but it is very important that you do it."
- Mahatma Gandhi

A. YOUR INTEREST IN COOLING SYSTEMS

Cooling devices have a fascination for people from a diversity of disciplines. Whether your interest lies in engineering realities or basic physics, at the manufacturer or consumer side, in down-to-earth diagnostics for malfunctioning hardware or establishing fundamental universal bounds for cooling performance from first principles, we believe this book has something to offer you.

Cooling systems permeate our daily lives and represent a substantial fraction of the world's total energy and power consumption, primarily household refrigerators, air-conditioning of buildings and industrial refrigeration. For conciseness we'll refer to all these applications by the simple engineering rubric "*chillers*" unless there is a specific need to distinguish among them.

The term "heat pump" describes a nominal cooling system where the useful effect extracted is the heating from heat rejection rather than the cooling from heat removal. The basic physics and engineering of heat pumps are qualitatively the same as for the corresponding cooling device.

Most of the material in this book is couched in the terms, nomenclature and variables of cooling systems. The application to heat pumps is straightforward, since each energetic flow and each source of irreversibility remains the same – only the useful effect changes. In order to strengthen these claims, we have included examples of how the analytic tools developed here can be applied specifically to heat pumps, with comparisons to real commercial devices. In the process, we will illustrate how systems that are specifically geared toward heating or temperature boosting are designed with a different balance of irreversibilities that is more favorable to higher temperature operation.

There is a diversity of interests in chillers. On the engineering side there are: (1) the manufacturers whose interest is to maximize thermodynamic efficiency subject to economic constraints; (2) the designers of cooling installations; (3) the chiller installers; (4) the engineers responsible for diagnostic and corrective measures; and (5) the consumers who pay the capital costs and energy bills. On the physics side there are researchers and students interested in: (*a*) the fundamental limits to the performance of cooling devices; (*b*) to what extent these bounds are device-independent; (*c*) how to bring these limits from idealized diagrams to the realities of commercial machines; (*d*) how actual chiller performance can be understood from basic irreversible thermodynamics and (*e*) how one can impose optimal controi strategies to attain maximum performance for a given technology.

B. COOLING BASICS

Cooling machines input power and transfer heat from a cold environment to a warmer one (Figure 1.1). They operate cyclically, namely, they continually repeat the same set of steps shown schematically in Figure 1.1, so that the working fluid of the device, called the refrigerant, returns to the same initial state for each cycle.

At a simplistic level, chillers can be viewed as heat engines operated in reverse, i.e., with the directional arrows for heat and work flows reversed. For example, whereas a heat engine produces power, a chiller inputs power. Whereas a heat engine accepts heat from a hot reservoir and rejects it to a cold reservoir, a chiller removes heat from the space to be cooled and rejects it to a warmer environment.

Air conditioning and refrigeration are the major applications toward which this book is geared; but it is not restricted to them. Rather, we will be developing general thermodynamic models for a wide variety of cooling devices and a broad range of operating conditions. So the applications can be whatever you find can benefit from these machines. We will carefully delineate the classes of chillers for which the analytic models developed here have been validated, and will establish the conditions under which the modeling tools we prescribe render accurate predictions.

In addition to a chiller's cooling rate (in kW), we will refer extensively to a dimensionless figure of merit called the **Coefficient of Performance** or COP for short. The COP is the ratio of the useful

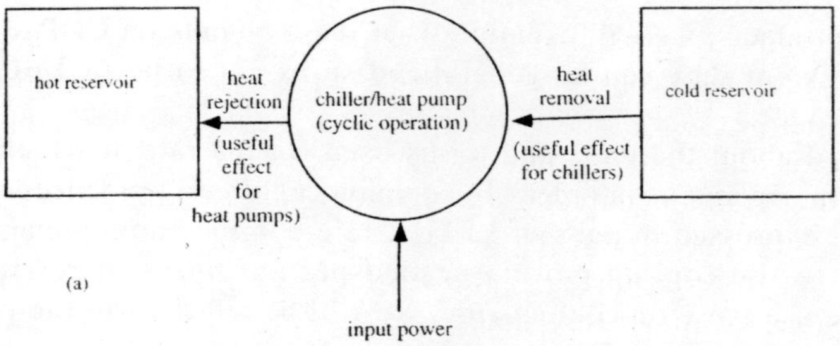

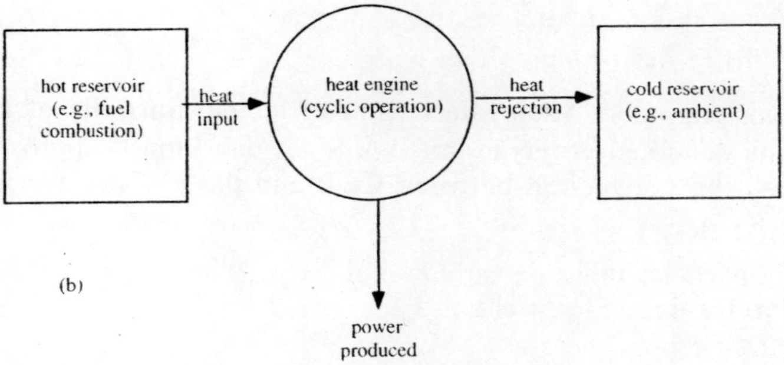

Figure 1.1: Schematic for cooling (chiller) and heat pump systems (1a), and for heat engines (1b). Power is input to cooling systems, which then remove heat from a cold reservoir and reject it to a hot reservoir. The process is cyclic and repeats continually, i.e., the refrigerant (working fluid of the system) returns to its initial state at the beginning of each cycle. Heat engines work in reverse, accepting heat from a hot source, producing power, and rejecting heat to a cold reservoir. The key distinction between chillers and heat pumps is where the useful effect is extracted: cooling (heat removal) at the cold side for chillers and heating (heat rejection) at the hot side for heat pumps.

effect produced to the input power. For example, for the common mechanical chiller,

$$\text{COP} = \frac{\text{cooling rate}}{\text{electric input power}}$$

where both numerator and denominator are expressed in the same units. Whereas cooling rate is limited by the size of chiller components, the COP is restricted by fundamental thermodynamic principles. While analyzing chiller models ranging from highly idealized to actual commercial products, we will examine what these bounds on COP are, and to what extent they can be generalized so as not to be tied to a particular device.

A word about the units and terms used for the rate at which cooling systems operate is in order. In common engineering practice, cooling *capacity*, expressed in units of kJ kg^{-1}, refers to the cooling energy (as opposed to the cooling power) needed per unit mass of refrigerant. When we use the term cooling *rate*, we will be referring to the product of cooling capacity and refrigerant mass flow rate:

$$\text{cooling rate} = \text{cooling capacity} * \text{refrigerant mass flow rate.}$$
$$\text{kW} = \text{kJ kg}^{-1} \qquad \text{kg s}^{-1}$$

Cooling engineers often rate chiller efficiency in units of kW per Rton (Rton denoting refrigeration tons). Since one Rton is approximately 3.517 kW, the conversion between COP and the kW per Rton rating is

$$\frac{\text{kW}}{\text{Rton}} = \frac{3.517}{\text{COP}}.$$

Most commercial chillers in the world are mechanical chillers, meaning that an electrically-driven mechanical compressor is used. The most common types are reciprocating, centrifugal and screw compressors, all of which are illustrated in Chapter 2 along with analyses of their relative advantages and limitations.

As illustrated schematically in Figure 1.2, the cycle starts by adiabatically compressing a refrigerant vapor in a mechanical compressor, thereby also increasing the vapor's temperature. In the condenser, the refrigerant rejects heat to the environment via a heat exchanger (a cooling tower) and exits as a liquid. The liquid is expanded in a throttler and enters the evaporator where heat is removed from the space to be cooled via a heat exchanger and boils the liquid. The emerging vapor is sucked into the compressor and the cycle is repeated.

Chillers can also be driven solely with heat, the most important example being absorption devices. Conceptually, they are similar to mechanical chillers, the key difference being that the role of a work-driven compressor is replaced by a heat-driven generator, as shown schematically in Figure 1.3. Heat input to the generator drives part of a volatile refrigerant out of a solution and into the vapor phase, with

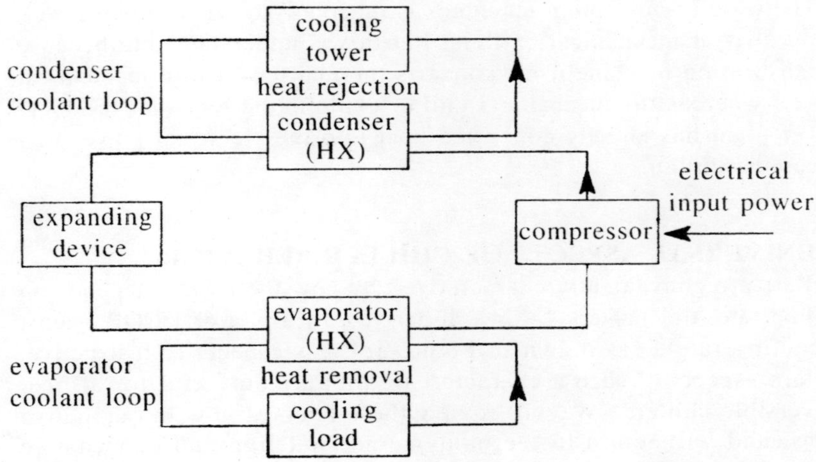

Figure 1.2: Schematic of a mechanical chiller. Heat transfers at the condenser and evaporator are effected through heat exchangers.

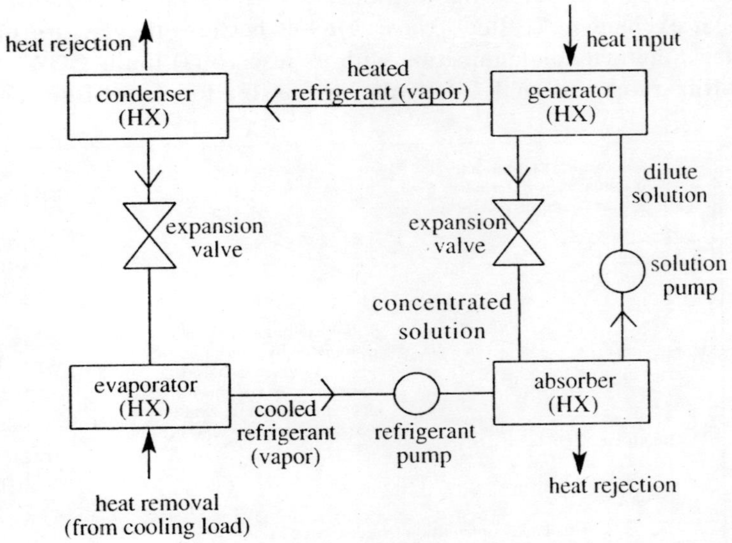

Figure 1.3: Schematic of an absorption chiller. Heat transfers at the generator, absorber, condenser and evaporator are effected through heat exchangers.

the reverse process carried out at the absorber. The condenser and evaporator serve the same functions as in mechanical chillers. A thorough discussion and illustration of how the cycle works are presented in Chapter 2.

The COP of absorption machines is inherently limited to being well below that of mechanical chillers; but this is understandable because the absorption machine itself converts thermal power into mechanical power, whereas the mechanical chiller exploits the fact that the local power plant has already converted heat into the electrical power that drives the chiller.

C. UNIVERSAL ASPECTS OF CHILLER BEHAVIOR

Chillers are conveniently characterized by how their COP depends on cooling rate. In Chapters 4–6, we'll derive why a plot of 1/COP against 1/(cooling rate) – as drawn in Figure 1.4 – is especially instructive. Certain aspects of such a characteristic plot are universal for all real irreversible chillers. We will review these features now in qualitative terms, and will return to thermodynamic modeling and quantitative observations in Chapters 4–6.

All real chillers appear to have irreversibilities that disfavor *both* high *and* low cooling rates. As an example, consider mechanical chillers. At high cooling rates, the bottleneck of finite-rate heat transfer in the heat exchangers (called *external* losses because they stem from the chiller's thermal communication with its reservoirs) limits COP. At low cooling rates, internal losses due to dissipation from fluid and

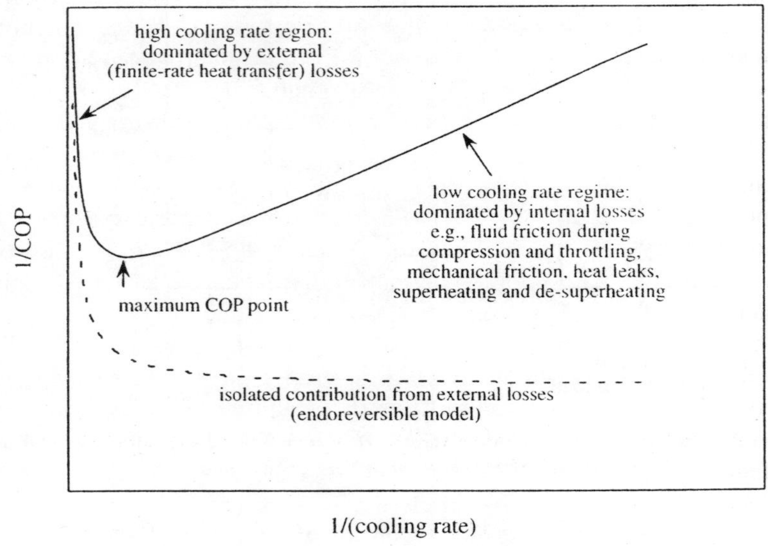

Figure 1.4: Characteristic chiller performance curve of 1/COP against 1/(cooling rate), drawn to illustrate qualitative trends. The endoreversible chiller, *i.e.*, the isolated contribution of external losses, corresponds to the broken curve.

mechanical friction, throttling, superheating and de-superheating govern COP. There is an intermediate range in which COP passes through a maximum.

The isolated contribution of external losses is shown by the broken curve in Figure 1.4, and is usually referred to as the endoreversible chiller model. ("Endoreversible" means internally reversible, namely, all losses are concentrated in the chiller's energetic exchanges with its surroundings.) In this limit of vanishingly small internal losses, the COP is maximized in the reversible limit of zero cooling rate. The inadequacies of the endoreversible model for *real* chillers are addressed at length in Chapter 11.

Note that whether external or internal losses dominate chiller performance is not necessarily a question of the physical speed of the chiller. Both types of losses are invariably present. And cooling rate need not correspond to the physical speed of the chiller. Exactly how chiller cooling rate is varied and its relation to the physical speed of the chiller will be considered in Chapter 2. The issue is the relative balance between external and internal losses, and how they affect the cooling rate dependence of the COP.

Chiller designers and manufacturers usually aim to have the maximum COP point occur at or near the machine's maximum cooling rate. In part, this is because properly-designed systems should run near their maximum cooling rate most of their operating time (since maximum capacity is a key variable for which one is paying). Hence for both the manufacturer and the consumer, accurately identifying the conditions of maximum COP is an important goal.

If maximum cooling rate should roughly coincide with maximum COP, then part-load chiller operation falls in the regime dominated by internal losses. Chiller performance data that we'll be analyzing in Chapters 4–10 will reinforce this basic fact of chiller design and operation. In those chapters, we'll also be showing how the parameters with which we can thermodynamically characterize a chiller can be extracted from performance plots in the form of Figure 1.4.

The specific irreversibilities noted above pertain to mechanical chillers. But as our analyses of a variety of chillers will disclose, the fact that there are always irreversibilities that disfavor both fast and slow cooling rates is independent of chiller type. This point will be documented thoroughly for absorption machines, and demonstrated for thermoelectric and thermoacoustic refrigerators as well.

D. OBJECTIVES OF THE BOOK AND THE INTENDED AUDIENCES

D1. The issues addressed and the predictions validated

We have tried to develop the analysis of cooling systems in a manner that can appeal to both the physicist and the engineer, and can form a bridge between the two communities in their analysis and presentation of cooling devices. A key question we will be answering is: are there universal elements in the thermodynamic performance of all chillers from which accurate predictive and diagnostic modeling tools can be developed? If so, to what degree would models based on these common elements be valid for refrigeration devices as diverse as mechanical, absorption and thermoelectric chillers?

Our approach is to capture the basic physics of the problem, and to emerge with quantitatively accurate predictive and diagnostic tools of substantial value to cooling engineers. We aim for thermodynamic models that are sufficiently simple that a chiller performance formula can be derived analytically, and the functional dependences of chiller performance on the major operating variables are transparent. The models will have to stand the test of comparison against experimental performance data.

We have found that in order to arrive at relatively simple analytic models, we needed to compromise certain aspects of detailed rigorous distributed thermodynamic modeling. Where appropriate, we will identify and explain these approximations. The reason we proceed with the approximate modeling procedures is the excellent agreement between model predictions or correlations and actual chiller performance data. By examining the nature of the inexactitudes, we hope the reader will also understand the limitations of these modeling procedures.

Beyond the issue of accurate model predictions, we focus upon accounting for the principal *trends* or qualitative features of chiller behavior. For example, referring to Figure 1.4, we see 3 key trends: (a) the decrease of COP with cooling rate in the regime dominated by external losses; (b) a roughly linear region in which COP increases with cooling rate as a consequence of internal dissipation; and (c) a point where COP is maximized at the optimal balance between these two distinct classes of irreversibilities. No *general* thermodynamic models for cooling devices have accounted for these trends. The reversible or Carnot limit of chiller behavior is simply a single point on such a plot (in fact, in the limit of zero cooling rate).

The modeling approaches developed in this book provide a connection between the universal reversible limit taught in all thermodynamics courses, and the real world of commercial chillers. Strictly rigorous models are, by their very nature, case-specific. By invoking reasonable approximations, we find that we can establish a sort of "base case" for chiller analysis. All the significant trends are accounted for. Fundamental limits on chiller COP as a function of practical operating variables *can* indeed be established and categorized.

The types of predictions we'll be making and testing are how chiller thermodynamic performance depends on: (1) cooling rate; (2) coolant (reservoir) temperatures; (3) coolant flow rate; (4) properties of the heat exchangers and how they are divided between the hot and cold sides of the chiller; (5) properties of the compressors and expansion devices in mechanical chillers; (6) generator and absorber characteristics for absorption chillers; (7) how the total time the refrigerant spends on one cycle is distributed among the assorted chiller components; and (8) an accurate accounting of entropy production and how it translates into the power required to drive the chiller.

Thermodynamic models for real chillers (as opposed to idealized unrealistic constructs) have tended to be case-specific. When accurate performance predictions are required over a wide range of cooling rates, many experiments must be performed, and extrapolation beyond the measured range may not be valid. It also means that diagnostic capabilities based upon a modest number of measurements are unfeasible.

The thermodynamic models developed in this book afford accurate predictions of chiller performance over a broad span of operating conditions from a handful of judiciously-chosen measurements, and can be used for rapid diagnostics. In addition, in capturing the basic physics of the irreversibilities that govern chiller behavior, these models provide a common framework for understanding and comparing the fundamental performance characteristics of all chillers: reciprocating, centrifugal, screw–compressor, absorption, thermoelectric, thermoacoustic or otherwise.

D2. The readership: toward whom the book is geared

We have several audiences in mind.

1) Cooling and air-conditioning engineers, and practitioners and researchers in the engineering sciences:

A central aim for this audience is to be able to characterize a cooling or refrigeration system with a relatively simple model that can be used for diagnostic and predictive purposes. Chillers are invariably complex machines. Detailed modeling of each chiller component, from first principles, is possible. But it represents a monumental task the results of which will probably be limited to the particular device under consideration. Massive simulations have been developed for these objectives, but usually are not practical tools for the analysis of installed cooling systems. For example, say you want to characterize an operating chiller by 2 or 3 parameters with which steady-state chiller performance can be predicted under a broad range of operating conditions (*e.g.*, environmental temperatures, coolant temperatures and flow rates, or cooling power demand). Equally important, if chiller performance degrades with time, you might want to identify the source of the problem, and to quantify the worsening of chiller efficiency or cooling rate under a spectrum of anticipated operating conditions. Or your interest may lie in measuring, *in situ*, how a given improvement you have devised has impacted chiller output.

Practicing engineers need this type of information for the design, monitoring and diagnosis of installed chillers. They also prefer that the model parameters be measurable non-intrusively. Namely, they need to relate to the chiller as a sort of blackbox, the internal properties of which must be probed with external measurements only.

The student and researcher will also expect that the thermodynamic models be physically transparent, namely, that model parameters have a distinct physical meaning linked to the characteristics of assorted chiller elements. Whereas the engineer in the field may suffice with completely empirical best-fit equations for expressing chiller performance, the student and researcher will demand a clear understanding of the processes involved, even if it comes at the level of lumping many complicated processes into only a few physically-meaningful variables. For example, compressors in mechanical chillers comprise many components, and their performance is determined by a combination of many complicated processes such as turbulent fluid flow, mechanical friction, fluid friction, the timing and placement of moving parts such as pistons or vanes, *etc.* Yet for purposes of predicting the performance of installed chillers, it is possible to characterize compressor performance solely in terms of the rate of entropy production. We'll be showing how this parameter can be determined experimentally with non-intrusive measurements.

The models developed, documented and verified in this book address the concerns of this audience. We show that for these purposes, the chiller can be treated as a sort of input-output device, viewed from the outside and probed only with externally-measurable parameters such as power input, cooling rate and coolant temperatures.

The chiller models derived here involve only 2 or 3 parameters. We delineate the link of these parameters to the particular irreversibilities that dictate chiller performance, specifically, internal dissipation, finite-rate heat exchange and heat leaks. After describing how these thermodynamic models can be used for both predictive and diagnostic purposes, we'll illustrate the power of these simple models with case studies based on commercial chillers and actual measured data.

The thermodynamic modeling developed here is applied to both mechanical and absorption chillers. The sub-division among mechanical chillers depending on the type of compressor used, *e.g.*, reciprocating, centrifugal and screw compressor, is also considered. The compressor type affects the range of cooling rates the chiller can traverse, the relative balance among the different sources of irreversibility, and hence the attainable efficiencies.

We will also show that although what has been referred to as an exergy (nominal Second Law) analysis for chillers may be of value to national energy planners and power plant designers, it is of little value to the key players in the chiller community: the consumers and the manufacturers. We will demonstrate how a correct Second Law analysis can be applied from the viewpoint of the consumer and manufacturer to arrive at optimized designs and performance criteria that can be noticeably different from those based on exergy analyses.

2) The basic scientist and students of the basic physics of cooling systems

2a) What are the universal principles that underlie the performance of thermodynamic cooling machines? Most of us are familiar with the fundamental limits for reversible machines, based on the First and Second Laws of Thermodynamics. But we also know that these limiting reversible efficiencies are far beyond the performance of even the most efficient state-of-the-art cooling systems. The role of irreversibilities is dominant and essential for modeling and understanding the problem.

Can one introduce irreversibilities and still emerge with a universal model? Alternatively put, can one derive comparable performance limits for real irreversible chillers? The models developed here, backed up with experimental data, take a step toward answering these questions.

Irreversibility (entropy production) is not an esoteric or arbitrarily-defined theoretical quantity. Rather, dissipation translates directly into performance variables such as cooling power and COP. The formalism that effects this translation, from both an engineering and a physics perspective, is developed in detail in Chapter 4.

2b) During the past 20 years, a large number of journal papers were published advocating endoreversible chiller models, *i.e.*, all the irreversibilities residing in the finite-rate heat exchange between the refrigerant and its reservoirs. Internal dissipation is viewed as negligible.

But, as we will demonstrate conclusively with chiller performance data, external losses are far from the full picture. The predominant loss mechanism in the vast majority of real chillers is internal dissipation, and ignoring it – excluding the key physics of the problem – results in predictions that not only grossly miss the mark quantitatively, but also fail to account for fundamental qualitative trends in chiller behavior. The so-called fundamental endoreversible limits on chiller performance are correct for the fictitious idealization assumed, but bear no resemblance to any real chiller. The papers advocating endoreversible models as representing real chillers suspiciously lack comparisons to the wealth of available experimental measurements that unequivocally attest to the inadequacy of endoreversible chiller models.

2c) In the fundamental chiller model of Chapters 4–6, the chiller parameters have a clear physical meaning. In fact, in order to powerfully establish this point, we present the exercise of first determining the magnitude of the 3 principal classes of irreversibility (internal dissipation, external heat exchange and heat leaks) indirectly from fitting chiller data to the models, and then intrusively and directly measuring these irreversibilities without regard to the model. The agreement between the two sets of results attests to the validity of assigning a particular physical significance to each model parameter.

2d) A pedagogical tool often used in helping the student to understand the thermodynamic performance of chillers is the temperature–entropy (*T–S*) diagram. The idealized reversible chiller cycle illustrated in introductory thermodynamics courses is comprised of rectangles on *T–S* plots, and the areas bear the simple interpretations of work input and cooling energy. Can this graphical representation be applied to real irreversible cycles and still retain the simplicity of rectangular elements and their physical interpretation? In Chapter 13, we'll show how this is accomplished. The method relies upon a careful analysis of entropy production and how it is translated into the work input required by the chiller.

The basic scientist may also be interested in far less common (but not necessarily less interesting) types of chillers, *e.g.*, thermoelectric and thermoacoustic refrigerators, which are also covered briefly in Chapters 2, 10 and 14. The thermoelectric refrigerator is an especially attractive illustration because it is the only cooling system of which we are aware where, simply by turning a dial (a rheostat), one can experimentally access the complete range of theoretically-realizable cooling capacities, from zero to the maximum cooling rate for a given device. For every value of attainable cooling rate, *two* values of COP are possible (at high and low electrical current). The degree to which the universal thermodynamic models developed for mechanical and absorption chillers can also be extended to these more exotic chiller types may be of interest to those in search of the universal principles that underlie the operation of real irreversible cooling systems.

A relatively new and intriguing direction is that of quantum-mechanical refrigerators, *i.e.*, molecular-level chillers. Their analysis is beyond the scope of this book; but we bring the reader's attention to this novel approach in [Bartana *et al* 1993; Geva & Kosloff 1996].

3) Chiller manufacturers

A primary interest of chiller manufacturers is to produce the best chiller at the lowest cost. In the course of in-house development and testing of a new chiller, the company needs tools for predicting and measuring how a given modification in a chiller component will affect COP and cooling rate. And if some unexpected change occurs because changing one element has an unanticipated indirect effect upon other components, the manufacturer needs a means for seeing that influence in the laboratory in terms of readily-measured variables. Also, the firm may wish to ascertain the combination of operating conditions of individual components that maximizes chiller efficiency at a given cooling rate: in short, the thermodynamic optimization of the chiller for a given investment.

To what degree has the empirical evolution of chiller design and construction reached truly optimal performance? We will show how thermodynamic modeling can be used to answer this query. For the best commercial chillers currently available, given the technological level in which the manufacturers have been willing to invest, we'll show that chiller performance is near the theoretical maximum.

4) University and industry courses

The material in this book can constitute part of a university course on cooling systems, or sections can be included in introductory and advanced thermodynamics courses. It represents a fundamental, relatively simple yet accurate modeling approach to a broad spectrum of real chillers. Both engineering-oriented and physics-oriented topics are covered in most of the chapters.

Many of the chapters here can serve as an industry-oriented course tailored to cooling engineers responsible for the installation, monitoring or diagnosis of chiller, refrigeration and heat pump units. In this spirit, we have suffused the book with examples rooted in *actual commercial* machines.

Selected chapters can be used in workshops for chiller design engineers at the companies that manufacture cooling equipment, both for in-house diagnostic testing and for the optimization of cooling hardware being developed. Chapters 2, 3, 4, 5, 6, 8 and 10 are tailored in part to this aim.

E. THE READER'S BACKGROUND

We assume the reader is familiar with:

a) *Elementary thermodynamics.* (Nonetheless we will review fundamental elements of the thermodynamics of cooling machines in Chapters 2 and 4.)

b) *How cooling loads are calculated.* We'll be focusing on chiller performance for a given known cooling demand. We will not be reviewing how one estimates the cooling requirements of a given office space or refrigeration plant. Reviews of the properties of air-water vapor mixtures and how they affect cooling loads, as well as descriptions of cooling towers and evaporators and how they operate, can be found in [Çengel & Boles 1989; Kreider & Rabl 1994].

c) *Basic thermal physics and engineering* (the rudimentary elements of heat transfer).

d) *Basic mathematical regression methods* (linear and multiple-linear regression).

We use metric units only, and have added a conversion table to facilitate conversions between metric and British units.

Chapter 2

THERMODYNAMIC AND OPERATIONAL FUNDAMENTALS

"Everything should be made as simple as possible, but not simpler." - Albert Einstein

A. INTRODUCTION

Although much of the thermodynamic modeling and analysis in this book relates to cooling systems effectively as blackboxes that must be characterized strictly from external non-intrusive measurements, it is important to have some appreciation of the contents of those blackboxes. What are their principal components? What types of thermodynamic cycles are involved? What are the fundamental limits on chiller or heat pump performance? What are the main irreversibilities? Where do these irreversibilities enter and how do they impact thermodynamic performance? Of what practical aspects of specific chiller components should the reader be aware prior to entering the realm of thermodynamic modeling? These are the issues we will try to address succinctly in this chapter.

The chapter divides primarily into the two most general categories of cooling devices: work-driven (mechanical) and heat-driven (absorption). At the end of the chapter we will also look at two non-conventional chillers, based on the thermoacoustic and thermoelectric effects.

We move from the general to the specific. First, we review the derivation of fundamental upper bounds for thermodynamic performance, with little regard to the particulars of the machine. The results are essentially device-independent. One would imagine that in designing real cooling systems, the properties of these idealized maximum-performance machines should be imitated to the greatest extent possible. The degree to which this can be accomplished is discussed, along with examples of the cooling cycles that have evolved as the preferences of the chiller industry. The derivations of actual performance equations for real chillers are reserved for Chapters 4 and 5.

B. MECHANICAL CHILLERS

B1. Reversible Carnot refrigeration cycle

A device-independent upper bound on chiller thermodynamic performance can be established by considering an idealized reversible thermodynamic cycle. Usually called a Carnot refrigeration cycle, it comprises 4 reversible branches, as portrayed in Figures 2.1 and 2.2:

1) Work W is input, adiabatically compressing the refrigerant and raising its temperature.

2) The refrigerant rejects heat Q_{hot} isothermally to a hot reservoir at temperature T_{hot}.

3) The refrigerant is expanded adiabatically.

4) Heat Q_{cold} is removed from the cold reservoir at temperature T_{cold} by isothermal transfer to the refrigerant.

The refrigerant then returns to the compression stage and the cycle is repeated. Because the compression and expansion branches are adiabatic and non-dissipative (*i.e.* isentropic), because all heat transfers are isothermal to or from an infinite reservoir, and because no loss mechanisms (irreversibilities) are introduced, the Carnot refrigeration cycle ensures that the maximum cooling energy is delivered (on branch 4) per unit of work input (on branch 1).

Since the cycle is reversible, it requires infinite time. That means that the average cooling rate and power input are zero. Furthermore, real heat transfer is driven across a non-zero temperature difference.

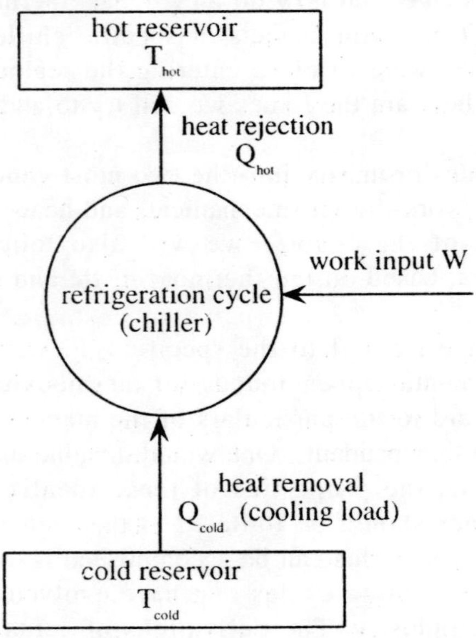

Figure 2.1: Schematic of the reversible Carnot refrigeration cycle.

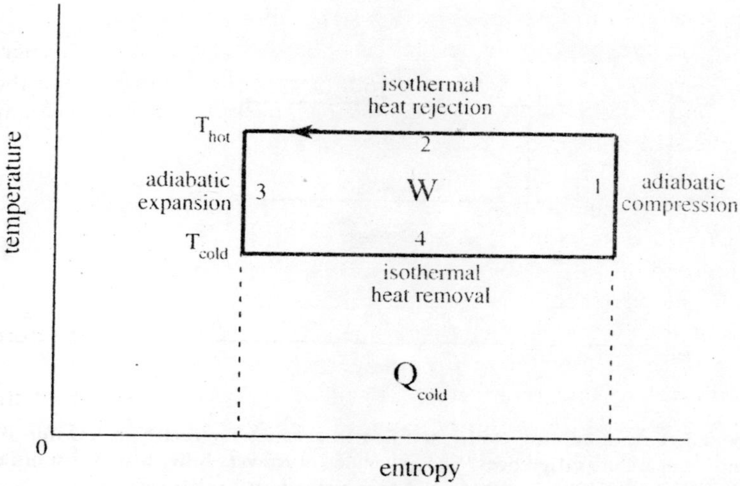

Figure 2.2: Temperature–entropy (T–S) plot for the Carnot refrigeration cycle. The heat rejection and heat removal branches are isothermal (horizontal lines), while the compression and expansion branches are isentropic (vertical lines). The area enclosed within the solid rectangle is the work input to the cycle, W. The area of the dotted-line (lower) rectangle is the cooling energy produced. Note that the direction for the refrigeration cycle is anti-clockwise, in contrast to the clockwise direction for heat engine operation.

That means that the reservoir temperature T_{hot} will be *below* the actual refrigerant temperature on the heat rejection branch T'_{hot}, and T_{cold} will fall *above* the actual refrigerant temperature on the heat removal side T'_{cold} (see Figure 2.3). The rates of heat transfer at the condenser Q_{cond} and evaporator Q_{evap} are proportional to the temperature differences $T'_{hot} - T_{hot}$ and $T_{cold} - T'_{cold}$, respectively.

Clearly, the reversible Carnot cooling cycle represents a highly idealized and limiting situation. The performance limit derived below is device-independent, just as the Carnot efficiency for heat engines is independent of how the heat engine may be constructed.

The figure of merit adopted in cooling engineering is the useful effect divided by the input power, defined as the Coefficient Of Performance, or COP for short. For chillers, the useful effect is Q_{cold} (branch 4). For heat pumps, the useful effect is Q_{hot} (branch 2).

$$\mathrm{COP}_{\mathrm{chiller}} = \frac{\text{cooling capacity}}{\text{work input}} = \frac{Q_{\mathrm{cold}}}{W} \tag{2.1}$$

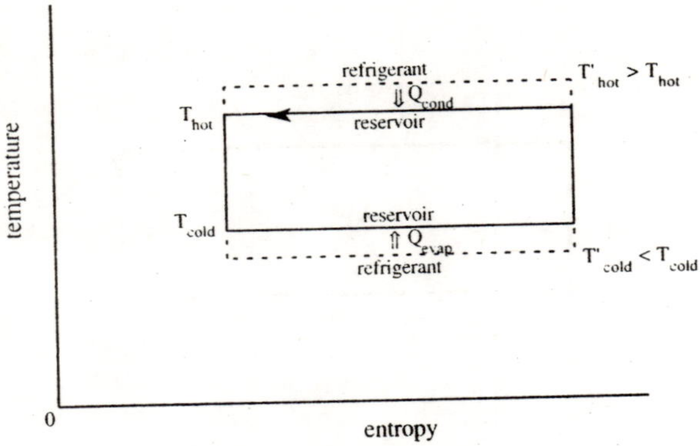

Figure 2.3: Carnot cooling cycle modified to account for real heat transfer across non-zero temperature differences. The only mode of irreversibility here is that of finite-rate heat transfer at the condenser and evaporator heat exchangers.

$$\mathrm{COP}_{\text{heat pump}} = \frac{\text{heat rejection}}{\text{work input}} = \frac{Q_{\text{hot}}}{W} \quad (2.2)$$

The fundamental upper bound on COP for the Carnot refrigeration cycle is derived as follows.

Recall that internal energy E and entropy S are state functions, so the change in their values for the refrigerant over one cycle at steady state is zero. Calculating the change over one cycle, we have

$$\Delta E = 0 = W - Q_{\text{hot}} + Q_{\text{cold}} \quad (2.3)$$

$$\Delta S = 0 = \frac{Q_{\text{hot}}}{T_{\text{hot}}} - \frac{Q_{\text{cold}}}{T_{\text{cold}}} \quad (2.4)$$

with all energy flows defined as positive. Combining Equations (2.1)–(2.4), we obtain

$$\mathrm{COP}_{\text{chiller}}^{\text{Carnot}} = \frac{T_{\text{cold}}}{T_{\text{hot}} - T_{\text{cold}}} \quad (2.5)$$

$$\mathrm{COP}_{\text{heat pump}}^{\text{Carnot}} = 1 + \mathrm{COP}_{\text{chiller}}^{\text{Carnot}} = \frac{T_{\text{hot}}}{T_{\text{hot}} - T_{\text{cold}}}. \quad (2.6)$$

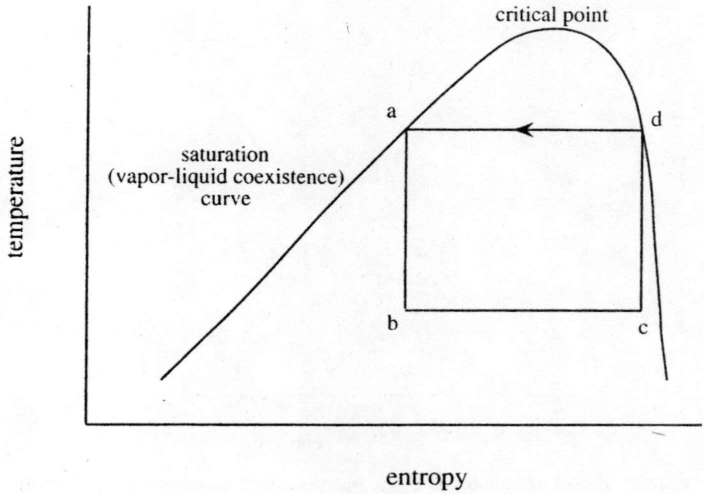

Figure 2.4: *T–S* diagram for the first possibility considered, wherein the cycle is fit within the refrigerant's saturation curve.

Reservoir temperatures T_{hot} and T_{cold} cannot in general be chosen at will. They are dictated by the application. For example, T_{hot} is usually ambient temperature and T_{cold} is commonly the temperature to be maintained in the cooled space.

B2. The discrepancy between physical idealizations and engineering realities

Because the Carnot cycle sketched in Figures 2.1 and 2.2 is the highest-COP cycle possible, one's first inclination is to try to mimic it as much as possible in real refrigeration cycles. The vapor-compression cycle is a natural choice because in principle the heat addition and heat rejection branches can be executed isothermally, at the phase transitions of evaporation and condensation. That might allow us to maintain the rectangular (Carnot-like) shape of the cycle on the *T–S* diagram.

The attractiveness of physical idealizations, however, is thwarted by engineering realities. Without entering into a myriad of mechanical complexities, let's try to understand why in basic physical terms.

A central problem is that compressors and throttlers (expansion devices) have difficulty efficiently handling two-phase mixtures. While the two-phase mixture problem on the compression and expansion branches could be overcome by operating outside the saturation region with a single phase, that would compromise maintaining isothermal

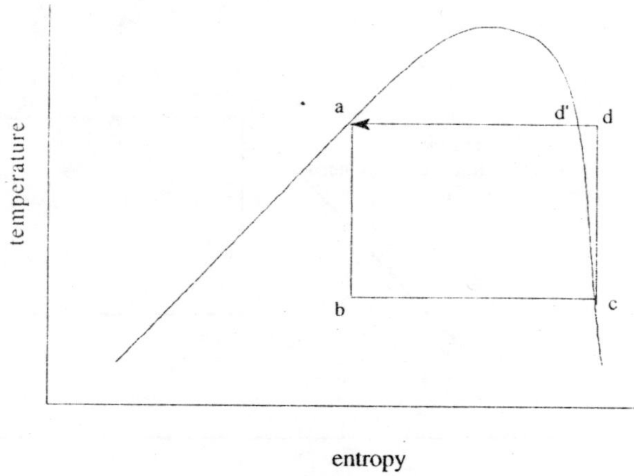

Figure 2.5: *T–S* diagram for the second possibility considered, wherein the cycle extends into the superheated vapor region. Branch *d–d'* is achieved via an isothermal compressor.

conditions for heat absorption and heat rejection. The vapor-compression cycles used in chillers represent the best compromise given material, economic and mechanical constraints.

To sharpen these arguments, let's consider a logical progression of 3 possibilities for fitting the Carnot cycle inside a vapor-compression machine, along with *T–S* diagrams to illustrate the points thermodynamically. The first attempt is represented in Figure 2.4, where the cycle is completely contained within the refrigerant's saturation (vapor–liquid coexistence) curve. A nice try, but impractical for 3 reasons. First is the problem of knowing when to terminate evaporation at point *c*, because no readily monitored variable such as pressure or temperature is changing along branch *b–c*. Second, the adiabatic compression branch *c–d* is complicated by a moist mixture at the compressor inlet. And third, the two-phase expansion *a–b* is quite difficult to achieve in a real (as opposed to an idealized) expander.

For our second and third options, we consider operating partly outside the saturation curve, in a single-phase region – in one case the single-phase region being the vapor, and in the other case the liquid. The former instance is illustrated in Figure 2.5. Two important problems that plagued the cycle of Figure 2.4 are resolved here: (*i*) point *c* can easily be sensed because the temperature starts to increase in an isobaric process as soon as the saturated vapor condition is met; and (*ii*) the adiabatic compression branch *c–d* is now approximately realizable so no liquid enters the compressor. Still, two key problems are: (1) expansion branch *a–b* remains impractical in real expansion devices,

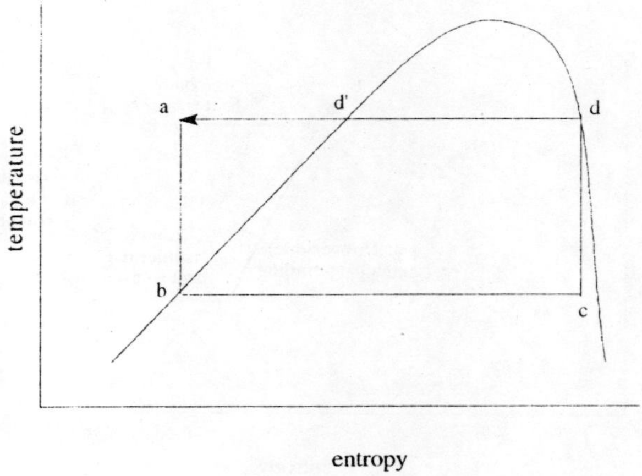

Figure 2.6: *T–S* diagram for the third possibility considered, wherein the cycle extends into the high-pressure liquid region.

as in Figure 2.4; and (2) the new additional isothermal compression branch *d–d'* requires an extra compressor and is difficult to realize with real equipment.

The third possibility is drawn in Figure 2.6. Its principal drawback is that point *a* becomes a very high pressure point (relative to the pressure at point *d'*), and renders the cycle impractical.

Before moving on to real chiller cycles, we consider the next logical pedagogical step: the idealized vapor-compression cycle illustrated in Figure 2.7. The four key steps are:

(1) throttling in an expansion device (*a–b*) during which the refrigerant temperature falls below the temperature of the space to be cooled;

(2) isobaric isothermal heat removal in the evaporator (*b–c*), with the refrigerant entering the evaporator as a low-quality saturated mixture and completely evaporates due to accepting heat from the refrigerated space;

(3) isentropic compression (*c–d*) where saturated vapor is brought up to the condenser pressure and well above the temperature of the surrounding medium; and

(4) isobaric heat rejection to the environment at the condenser (*d–d'–a*) of which branch *d'–a* is isothermal, with the refrigerant entering as superheated vapor and leaving as saturated liquid.

In principle, the problems noted above for the Carnot cycles considered in Figures 2.4–2.6 are overcome. At point *c*, the refrigerant exits the evaporator, and hence is sucked into the compressor, as dry saturated (single-phase) vapor at the evaporator pressure. At point *a*,

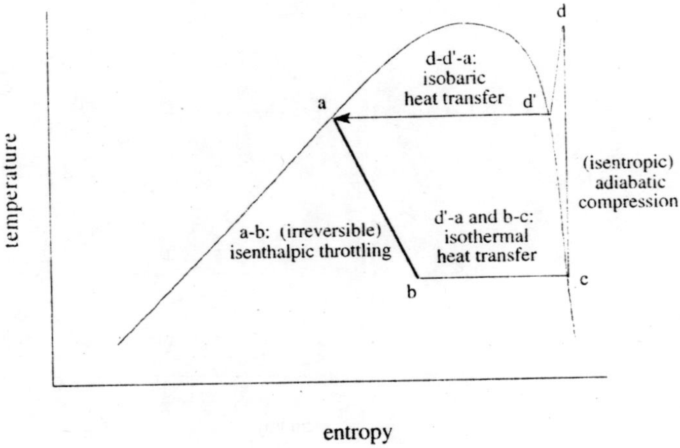

Figure 2.7: *T–S* diagram for the idealized vapor-compression cycle.

the refrigerant exits the condenser as a (single-phase) saturated liquid. Heat transfer branches *b–c* and *d'–a* can in principle be isothermal. The adiabatic compression branch *c–d* can in principle be isentropic, and proceeds all the way to the condenser pressure.

The dry compression and superheating along *c–d* causes the cycle to lose its rectangular shape on the *T–S* plot. The area that lies above the condensing temperature, often called the "superheat horn", represents additional work associated with dry compression, and hence a reduced COP.

To work with practical devices, we introduce a simple throttling valve for the adiabatic expansion branch *a–b*. This now becomes a constant-enthalpy, and not an isentropic (although still adiabatic), process. In other words, an unavoidable irreversibility is knowingly introduced (as well as another loss of the rectangular shape of the cycle on the *T–S* diagram). Were expansion to be executed isentropically, the resulting work would be exploited to help run the compressor. The introduction of an expansion engine is possible, but practical and economic factors mitigate against it, specifically: (*a*) the exploitable work is only a small fraction of that required by the compressor; (*b*) there are practical difficulties in using a two-phase mixture in the engine; and (*c*) the relatively high cost of this measure has not been commensurate with the savings.

Referring to Figure 2.7, we can express the chiller's cooling capacity and COP in terms of the refrigerant's specific (per unit mass) enthalpy h at different points along the cycle. Specifically (and recalling that $h_b = h_a$ for the isenthalpic throttling)

$$\text{cooling capacity} = h_c - h_b = h_c - h_a \tag{2.7}$$

$$\text{input work} = h_d - h_c \tag{2.8}$$

$$\text{COP} = \frac{h_c - h_a}{h_d - h_c}. \tag{2.9}$$

Tutorial 2.1

An idealized (theoretical) vapor-compression refrigeration cycle with ammonia as the refrigerant operates with a refrigerant condensing temperature of T_{cond} = 40°C and a refrigerant evaporating temperature of T_{evap} = −20°C. Compare the theoretical cycle with the corresponding Carnot cycle in terms of: (*a*) work input *W*; (*b*) cooling capacity Q_{evap}; and (*c*) COP. Consult standard thermodynamic tables [Mayhew & Rogers 1971, ASHRAE 1998] for the thermodynamic properties of ammonia.

Solution: Refer to Figure 2.8, which is adapted from Figure 2.7 for this problem, and identifies the states referred to in the calculations that follow. The standard thermodynamic tables consulted refer the specific entropy, *s*, of liquid ammonia to its value at −40°C, *i.e.*, $s_{liquid}(-40°C) \equiv 0$.

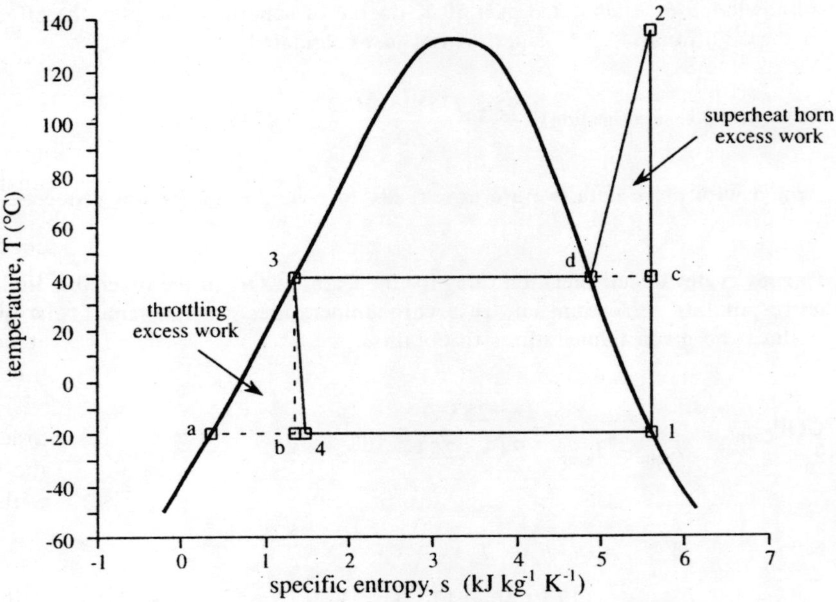

Figure 2.8: *T–S* diagram for the idealized ammonia vapor-compression cycle.

From standard thermodynamic tables [Mayhew & Rogers 1971, ASHRAE 1998], we find the following specific (per unit mass) properties for ammonia:

at −20°C	$h_1 = 1420$ kJ kg^{-1}	$s_1 = s_2 = 5.623$ kJ kg^{-1} K^{-1}
	$h_a = 89.8$ kJ kg^{-1}	$s_a = 0.368$ kJ kg^{-1} K^{-1}
at 40°C	$h_d = 1473.3$ kJ kg^{-1}	$s_d = 4.877$ kJ kg^{-1} K^{-1}.

We also note that $h_3 = 371.9$ kJ kg^{-1} and $s_3 = s_b = 1.360$ kJ kg^{-1} K^{-1}.

The values of s provided at 50 and 100 K degrees of superheating are:

s (at 50 K superheating) = 5.321 kJ kg^{-1} K^{-1}
s (at 100 K superheating) = 5.655 kJ kg^{-1} K^{-1}.

The actual degree of superheating is calculated by interpolation between values available in the tables:

$$\text{degree of superheating} = 50 + \frac{(100-50)\,(5.623-5.321)}{5.655-5.321} = 95.2 \text{ K}$$

i.e., 45.2 K above the tabulated value of 50 K superheating.

The specific enthalpy at the top of the superheat horn h_2 is obtained from a knowledge of h tabulated at a 50 K degree of superheating, plus the effect of the additional 45.2 K superheating just calculated:

$$h_2 = h(\text{at 50 K superheating}) + \frac{45.2(1751.9-1622.4)}{50} = 1739.46 \text{ kJ kg}^{-1}.$$

Armed with these data, we are now ready to generate the figures requested.

Carnot cycle: Recall Equation (2.5) for the Carnot COP. In the reversible limit, corresponding refrigerant and reservoir temperatures are the same. Now introduce the given temperatures to obtain

$$\text{COP}_{\text{Carnot}} = \frac{T_{\text{evap}}}{T_{\text{cond}} - T_{\text{evap}}} = \frac{253}{313-253} = 4.216.$$

Next,
$$\begin{aligned} W_{\text{rev}} &= (h_2 - h_d) - T_3\,(s_c - s_d) \\ &= (1739.46 - 1473.3) - 319\,(5.623 - 4.877) \\ &= 28.216 \text{ kJ kg}^{-1}. \end{aligned}$$

Finally, $Q_{evap}^{rev} = T_3 (s_1 - s_3) = 253 (5.623 - 1.360) = 1078.54$ kJ kg^{-1}.

As a quick consistency check for COP: $\dfrac{Q_{\text{evap}}^{\text{rev}}}{W_{\text{rev}}} = 4.216$, which of course is the same as the Carnot COP calculated at the beginning of the exercise.

Idealized vapor–compression (theoretical) cycle: Relative to the reversible Carnot cycle, this irreversible cycle incurs excess work due to: (1) the superheat horn, and (2) throttling. In addition it suffers from a loss of cooling capacity. Each of these is readily calculated as follows.

Excess work due to the superheat horn $= (h_2 - h_d) - T_3 (s_c - s_d)$

$= (1739.46 - 1473.3) -$

$319 (5.623 - 4.877)$

$= 28.216$ kJ kg^{-1}.

Excess work from throttling $= (h_3 - h_a) - T_4 (s_b - s_a)$

$= (371.9 - 89.8) - 253 (1.360 - 0.368)$

$= 31.12$ kJ kg^{-1}.

Loss of cooling capacity $= h_4 - h_b = T_4 (s_4 - s_b)$.

We take advantage of the following 3 thermodynamic equivalences for this particular cycle

$h_3 = h_4$ (since throttling is isenthalpic)

$s_b = s_3$ and

$h_b - h_a = T_4 (s_b - s_a)$

to obtain that the loss of cooling capacity $= (h_4 - h_a) - T_4 (s_3 - s_a)$

$= (371.9 - 89.8) - 253 \times$

$(1.360 - 0.368) = 31.12$ kJ kg^{-1}.

The irreversible cycle's COP can now be expressed in terms of the values of Q_{evap} and W for the reversible cycle, modified by the losses just calculated:

$$\text{COP} = \frac{Q_{\text{evap}}^{\text{rev}} - (\text{loss of cooling capacity})}{W_{\text{rev}} + (\text{two excess work contributions})} = \frac{1078.54 - 31.12}{251.78 + 31.12 + 28.216} = 3.324$$

which is 79% of the Carnot COP. This exceptionally high figure derives from the idealized nature of the irreversible cycle, *i.e.*, key loss mechanisms having been omitted.

(At this stage, Q_{evap} refers to cooling *capacity* in kJ kg^{-1}. In later chapters, where we will be examining cooling *power*, as the product of cooling capacity and refrigerant mass flow rate, we will also be using the symbol Q_{evap} to denote cooling *rates* in kW.)

Gas cycles (*e.g.*, reverse Brayton and reverse Otto cycles with isobaric or isochoric heat transfer) are possible, and in fact constitute a small fraction of special-application refrigeration systems. They suffer, however, from the heat transfer branches being substantially non-isothermal. This translates into a considerable increase in the required work input for the same cooling capacity, in other words, lower COP.

Other factors militate against adopting the idealized processes in the Carnot refrigeration cycle. For example, cycles must be performed in *finite* time in order to attain non-zero cooling rates. Heat exchangers must be finite in extent, so a thermal bottleneck develops in removing and rejecting heat. Compressors and throttlers incur significant fluid friction losses. Losses derive from superheating in the evaporator and de-superheating in the condenser (to ensure pure vapor or pure liquid at their exits). There are also heat leaks in each component and in refrigerant lines, as well as pressure drops, and mechanical friction losses in the compressor shaft. Hence actual COPs are far below the Carnot limit (a point to which we'll return at the beginning of Chapter 4).

B3. Real vapor-compression cycles

Figure 2.9 is a schematic of a real vapor compression chiller. The branches of the cycle are depicted in Figure 2.10:

- Two-phase refrigerant exits the evaporator and is superheated prior to being sucked into the compressor (1–2).
- Refrigerant vapor is compressed and discharged to the condenser (2–3).
- De-superheating in the condenser (3–4).
- Condensation/heat rejection (4–5–6).
- Throttling (expansion) (6–7).
- Evaporation/cooling effect/heat removal (7–1).

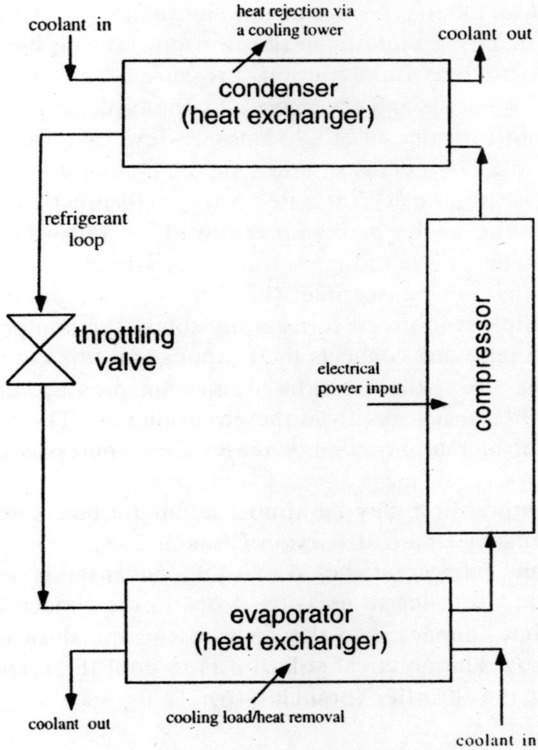

Figure 2.9: Schematic of a real vapor-compression mechanical chiller.

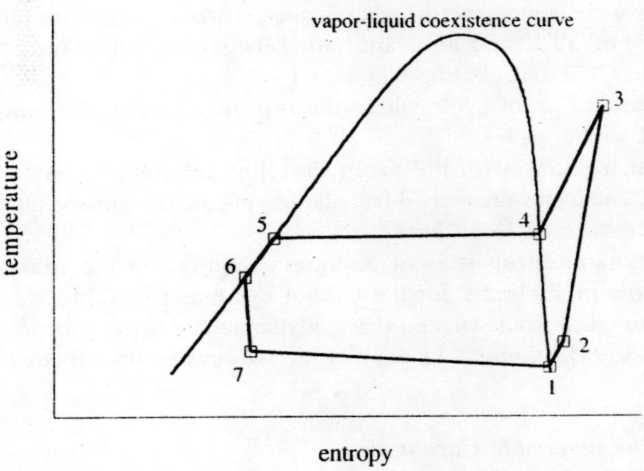

Figure 2.10: *T–S* diagram for a real (as opposed to an idealized) vapor-compression cycle.

What should ideally be isothermal branches (4–5 and 7–1) and isentropic branches deviate from the desired limiting behavior due to internal losses such as fluid friction (pressure drops), heat leaks to or from the environment, and the need for single-phase processes at the compressor and throttler. Let's examine a few specifics.

- Ideally, the refrigerant should exit the evaporator and be sucked into the compressor as dry saturated vapor. In practice, though, this is impractical due to the precision required in controlling the refrigerant's state. The pragmatic compromise is slightly superheating the refrigerant at the compressor inlet (branch 1–2), which guarantees that only a completely vaporized refrigerant enters the compressor.
- With the pipe that connects the evaporator to the compressor typically not being short, there can be significant pressure drops as well as non-negligible heat leaks from the environment. The associated increase in the refrigerant's volume increases the compressor's power input requirement.
- Real compression may be almost adiabatic but is not isentropic due to fluid and mechanical friction (branch 2–3).
- Ideally, the refrigerant should exit the condenser as a single-phase saturated liquid. But due to pressure drops in the condenser itself and in the lines that connect it to the compressor and throttler, this may not be the case. The practical solution is to cool the refrigerant prior to its entering the throttler (branch 5–6).

Tutorial 2.2:

A real refrigerator with ammonia refrigerant operates between a condenser refrigerant temperature T_{cond} of 32°C and an evaporator refrigerant temperature T_{evap} of –18°C. Find the COP and cooling capacity for:

(*a*) the reversible Carnot cycle where the refrigerant exits the compressor as dry saturated vapor; and

(*b*) the actual machine with the vapor entering the compressor at a temperature of –20°C and a pressure of 1.74 bar, and leaving the compressor superheated to 136°C at a pressure of 13.89 bar.

The thermodynamic properties of ammonia required in the analyses are summarized below in Table 2.1 for the Carnot cycle and in Table 2.2 for the real cycle, and are taken from standard thermodynamic tables [Mayhew & Rogers 1971, ASHRAE 1998]. Figure 2.11 depicts the two cycles investigated in this tutorial.

Solution: (*a*) *The reversible Carnot cycle*

The reversible limit corresponds to: (1) infinite heat exchanger thermal conductances for both the condenser and evaporator; (2) isentropic expansion and

Table 2.1: Thermodynamic properties for ammonia as required for the Carnot cycle

state point (see Fig. 2.11)	pressure (bar)	temperature (°C)	specific enthalpy (kJ kg^{-1})	specific entropy (kJ kg^{-1} K^{-1})
1_c	2.077	–18	1261.7	4.962
2_c	12.37	32	1469.9	4.962
3_c	12.37	32	332.8	1.235
4_c	2.077	–18	310.6	1.235

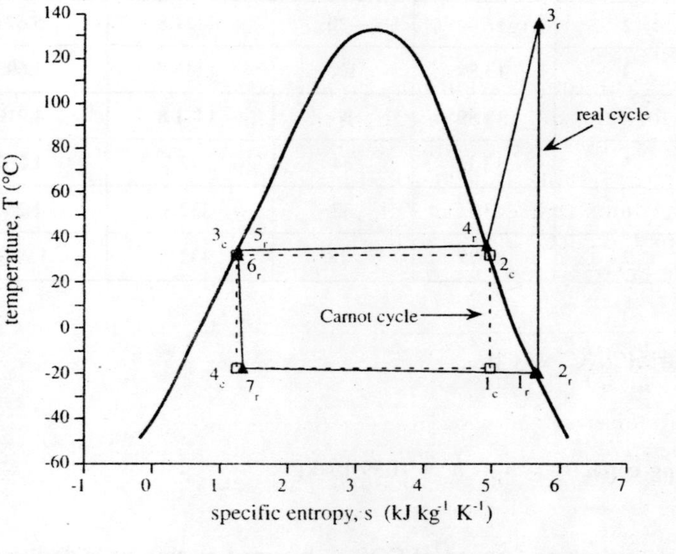

Figure 2.11: The *T–S* diagram for the Carnot cycle (subscript c, open squares) and for the real cycle (subscript r, solid triangles) of the ammonia–refrigerant machine considered here.

compression; and (3) no pressure drops or related losses for the refrigerant. The particular case where the refrigerant exits the compressor as dry saturated vapor corresponds to Figure 2.4 above, and is drawn as the broken-line rectangle in Figure 2.11.

$$COP_{Carnot} = \frac{h_1 - h_4}{(h_2 - h_1) - (h_3 - h_4)} = \frac{1261.7 - 310.6}{(1469.9 - 1261.7) - (332.8 - 310.6)} = 5.11$$

Cooling capacity $= h_1 - h_4 = 951.1$ kJ kg^{-1}.

(b) The real cycle
The actual irreversible cycle includes: (1) finite-size heat exchangers and hence finite-rate heat transfer losses; (2) non-isentropic expansion; (3) expansion via throttling; and (4) pressure losses incurred mainly in the single-phase flow region of the heat exchangers.

Table 2.2: The requisite thermodynamic properties of ammonia for the real cycle.

state point (see Fig. 2.11)	pressure (bar)	temperature (°C)	specific enthalpy (kJ kg^{-1})	specific entropy (kJ kg^{-1} K^{-1})
1_r	1.902	–20	1420.0	5.623
2_r	1.740	–20	1421.8	5.671
3_r	13.89	136	1745.7	5.692
4_r	13.89	36	1471.8	4.919
5_r	13.11	34	342.5	1.267
6_r	13.11	32	332.8	1.267
7_r	2.077	–18	332.8	1.3208

$$\mathrm{COP} = \frac{h_{2r} - h_{7r}}{h_{3r} - h_{2r}} = 3.36$$

Cooling capacity $= h_1 - h_4 = 1089$ kJ kg^{-1}.

The relatively high fraction of $\mathrm{COP}_{\mathrm{Carnot}}$ attained by this cycle derives from our not having introduced unduly small finite-rate heat transfer losses in the heat exchangers in this exercise, *i.e.*, our having implicitly assumed unusually large heat exchangers for common commercial chillers.

The condenser heat exchanger, evaporator heat exchanger and throttler (expansion device) are common to all mechanical chillers. Construction and design details appear in standard texts such as [Stoecker & Jones 1982; Kreider & Rabl 1994]. The type of compressor used is what distinguishes among the classes of mechanical chillers. The particulars of the main compressor categories, as they relate to the the predictive, diagnostic and optimization studies of concern in this book, will now be reviewed briefly.

B4. Reciprocating Chillers

Reciprocating chillers (see Figures 2.12 and 2.13) represent the lion's share of installed cooling capacity in the world. Typically, reciprocating chillers are used for cooling loads ranging from a fraction of a kW up to about 300 kW. Each cylinder has a moving piston and suction and discharge valves. In contrast, centrifugal and screw compressors employ rotating elements. Reciprocating compressors can be single- or multi-cylinder.

When cooling loads below maximum or rated capacity are required, there are two ways in which a reciprocating chiller responds. The range down to around 70% of full capacity can be realized by changing the coolant temperatures, which in turn affect the state of the refrigerant as it enters the compressor, throttler and heat exchangers. The compressor continues to operate at a fixed number of cycles per second. All cylinders are active (loaded). In this instance, low cooling rates are not linked to slow physical operation. We emphasize this point at this juncture, because in later chapters we will invoke experimental measurements of reciprocating chillers to demonstrate that at *low* cooling rates internal losses that stem from processes such as fluid and mechanical friction dominate the performance of real chillers. There is

Fig. 2.12: Photograph of a reciprocating chiller.

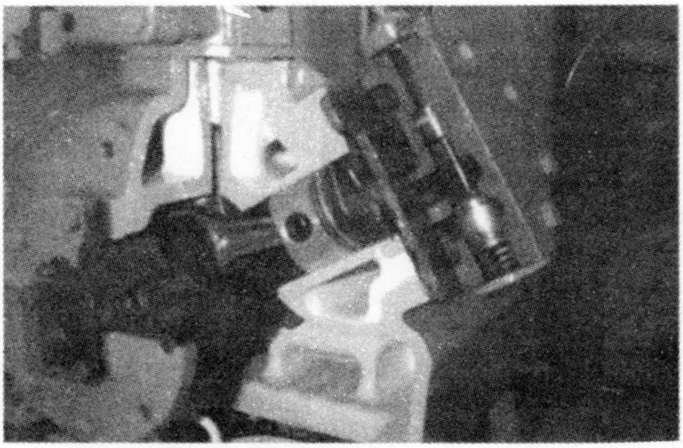

Figure 2.13: Photograph of a reciprocating compressor with its valve plate and suction valves. Refrigerant enters through a valve by suction. The valve closes and the piston compresses the refrigerant, until a second valve (on the opposite side of the valve plate) opens to release compressed higher-pressure higher-temperature refrigerant.

no physical inconsistency between this observation and the fact that frictional losses grow rapidly with fluid or mechanical speed. Namely, low cooling rates do not necessarily imply slow compressor speeds.

For part-loads below around 70%, cylinders are unloaded, *i.e.*, rendered passive by leaving their valves open, while not causing a mechanical imbalance on the compressor linkage. The analyses developed in this book relate primarily to achieving part-load by varying coolant temperatures. In large chiller plants, low part-loads are readily attained by installing several small individual chillers of varying capacities, and turning one or more of those individual units off as needed. This tends to be a more energy-efficient method, although often at a greater capital investment.

B5. Centrifugal chillers

For relatively large cooling loads, starting at around 500 kW, centrifugal compressors (Figure 2.14) are the usual choice. The compressor design is akin to a centrifugal pump, with fluid entering at the center of the impeller and compressed to its edge by centrifugal force (Figure 2.15). Centrifugal compressors are made with only one wheel for low pressure ratios, but are generally multi-stage.

Centrifugal chillers are usually built to operate over a narrow range of coolant temperatures (evaporator coolant outlet temperature T_{evap}^{out} and condenser coolant inlet temperature T_{cond}^{in}), typically within ±2°C, even though they can supply cooling rates well below rated capacity. The inlet guide vanes in the centrifugal compressor are closed, to differ-

Figure 2.14: Photograph of a centrifugal chiller.

ing extents, to restrict the flow of refrigerant and thereby lessen cooling rate (irrespective of the number of stages in the chiller). Most centrifugal chillers can produce down to 30% of maximum capacity, and a small minority with special compressor designs can go as low as 15%. At cooling rates below around 30–40% of the maximum, the compressor blades may start to shake due to the aerodynamic phenomenon called stalling. In fact, most centrifugal chillers are equipped with a variable diffuser to avoid stalling even at part loads as high as 60% of the maximum. Accordingly, the consumer or designer should be wary of extrapolating manufacturer performance data below part loads of around 30% of maximum.

Typical building air-conditioning requirements for water-cooled evaporators demand that water temperatures in the cooling coils are supplied in the range 5–8°C for relatively good dehumidification, particularly for hot and humid climates. T^{in}_{cond} can change with the load, but will remain above the local wet-bulb temperature. However, in climates where the local wet-bulb temperature is as low as 15°C, it is customary in centrifugal chillers to include a condenser water bypass to boost T^{in}_{cond} so that the chiller can function properly. In essence, chiller manufacturers configure centrifugal chillers to maintain a fairly constant T^{in}_{cond}.

B6. Screw compressor chillers

For the intermediate cooling load range of 300–500 kW, screw compressors are often used (Figures 2.16 and 2.17). Their virtue relative

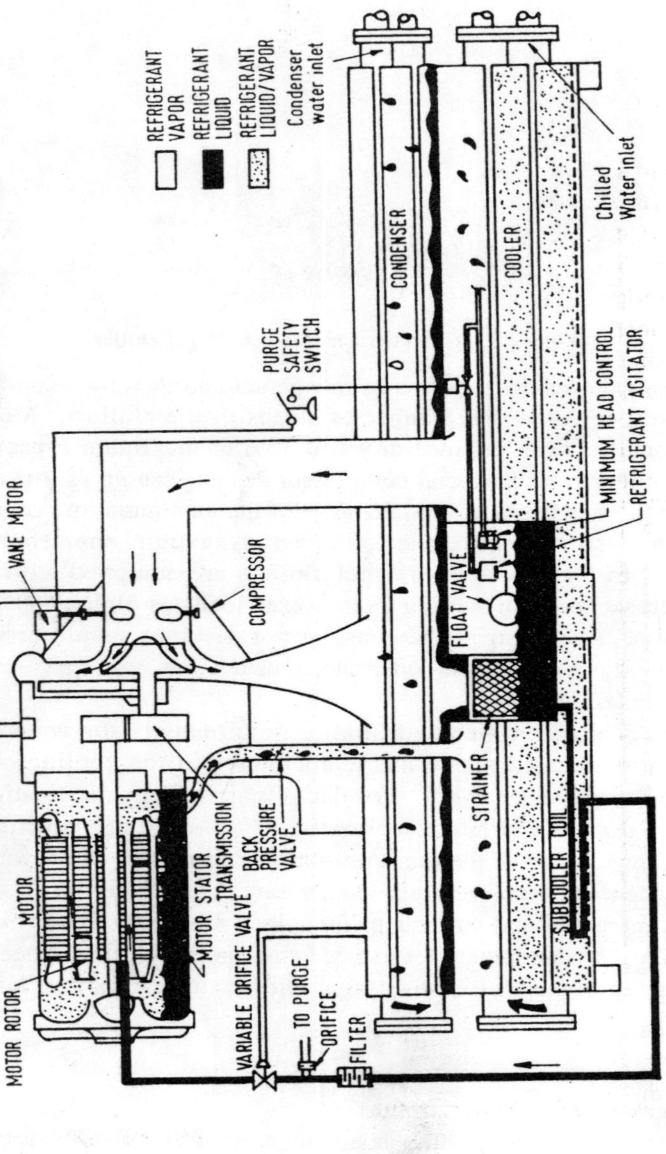

Figure 2.15: Schematic of the centrifugal compressor.

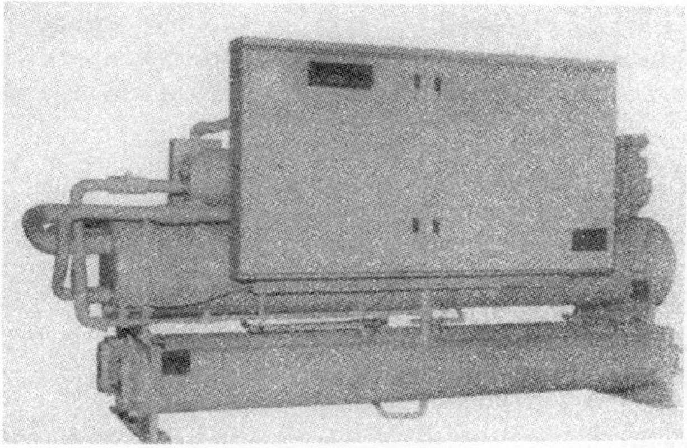

Figure 2.16: Photograph of a screw-compressor chiller.

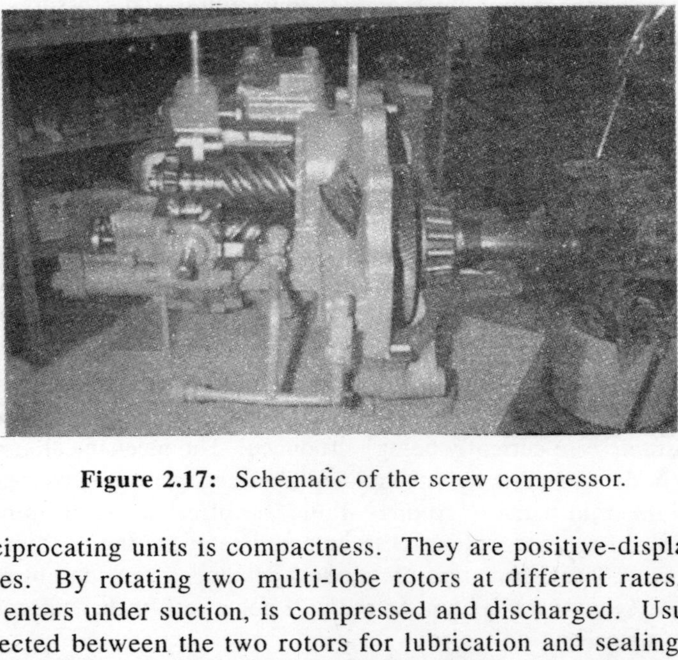

Figure 2.17: Schematic of the screw compressor.

to reciprocating units is compactness. They are positive-displacement devices. By rotating two multi-lobe rotors at different rates, refrigerant enters under suction, is compressed and discharged. Usually oil is injected between the two rotors for lubrication and sealing.

Screw compressors realize part-load conditions by unloading a sliding-valve unit which varies the compression ratio of the compressor. The variation in compression ratio in turn changes the refrigerant temperature in the condenser. Internal dissipation in the screw compressor stems primarily from refrigerant leakages between the rotors (*i.e.*, refrigerant leaks back to the suction port via the clearances of the rotors and lobes), especially at high pressure ratios. The compression characteristics of the screw compressor can be superior (and its internal dissipation lower) than those of the centrifugal compressor because the former can intake slightly wet refrigerant. Furthermore, screw compressors can safely produce part-load cooling rates as low as about 10% of maximum.

For efficient chiller operation, screw compressors are useful only for small thermal lifts (the difference between the condenser and evaporator refrigerant temperatures $T_{cond} - T_{evap}$) of around 20 K. In this range, the pressure drop is correspondingly low and leakage losses are lessened. Measured COPs as high as 6 are reasonable. But when the thermal lift increases to the range of 40–50 K, rotor leakages increase considerably and COP drops dramatically to around 1.

B7. Refrigerants

Refrigerant refers to the working fluid in the chiller. A refrigerant is selected in accordance with the extreme temperatures and pressures that must be accommodated for a given cooling need, *i.e.*, at the hot and cold reservoirs. Among the important desirable properties of a refrigerant are: (1) a high heat of vaporization, to achieve a large cooling capacity; (2) a low freezing point to avoid freezing at the low-temperature end of the cycle under extreme conditions; (3) a high critical point, to lower the required input power at the compressor; (4) an evaporation pressure of at least atmospheric pressure, to prevent air from leaking into the system; (5) a low condensation pressure, to avoid the need for expensive piping and equipment; (6) the chemical traits of non-toxicity, non-corrosiveness, non-flammability and chemical stability; and (7) low cost.

Until recently, refrigerants in mechanical chillers have predominantly been fluorinated hydrocarbons (CFCs); but due to environmental concerns, substitutes are currently being introduced. The mechanical chillers examined in detail in this book contained CFC refrigerants (often going by the commercial name of freons). Table 2.3 offers a list of common commercial refrigerants, along with selected relevant thermodynamic characteristics. The thermodynamic models developed here do not relate to a specific type of refrigerant; so we will not dwell further upon their material properties.

In typical vapor-compression (mechanical) chiller cycles, refrigerants may not be approximated as ideal gases. Consequently, one must refer to standard tables in which refrigerant thermodynamic properties such as enthalpy, internal energy and entropy as functions of temperature and pressure are tabulated in order to calculate the principal performance variables for each stage along the cycle. Such exercises are commonly reviewed in engineering texts, such as [Çengel & Boles 1989; Kreider & Rabl 1994].

Table 2.3: Physical properties of selected commercial refrigerants

refrigerant	molecular mass (kg kmole^{-1})	boiling point at p = 101.325 kPa (°C)	freezing point (°C)	critical temperature (°C)	critical pressure (kPa)	critical volume (l kg^{-1})	latent heat of vaporization (kJ kg^{-1})
CO_2	44.01	−78.4	−56.6	31.1	7372	2.135	230.54
chlorodifluoro-methane (R22) $CHClF_2$	86.48	−40.76	−160	96.0	4974	1.904	204.87
R–502 (mixture by weight of 48.8% R22 and 51.2% R115 chloropentane–fluoroethane $CClF_2CF_3$)	111.63	−45.5	---	82.2	4075	1.785	146.63
NH_3 (R717)	17.03	−33.3	−77.7	133.0	11417	4.245	1261.81
dichlorodifluoro-methane (R12) CCl_2F_2	120.93	−29.74	−158	112.0	4113	1.792	152.68
tetrafluoroethane (R134a) CF_3CH_2F	102.03	−26.16	−96.6	101.0	4067	1.181	198.68
trichlorotrifluoro–ethane (R113) CCl_2FCClF_2	187.39	47.57	−35	214.4	3437	1.736	157.97
H_2O	18.02	100	0	373.99	22064	3.11	2500.5

C. ABSORPTION CHILLERS

C1. Absorption basics and absorption versus mechanical chillers

Absorption cycles are similar to mechanical-chiller cycles in utilizing a condenser, evaporator and expansion device (see the schematic in Figure 2.18). The difference lies in how the low-pressure vapor that exits the evaporator is converted into the high-pressure vapor that enters the condenser. Instead of the work-driven compressor of a mechanical chiller, thermal power is the driving force. The heat is usually delivered in the form of hot water or steam, and is commonly derived from the combustion of natural gas, industrial waste heat, geothermal sources, or solar energy collection.

A vapor-compression chiller produces its cooling at an evaporator (a heat pump produces its heating at a condenser). The corresponding absorption system includes two additional heat reservoirs: a generator and an absorber. A volatile working fluid (refrigerant) is partially separated from the carrier solution by the heat input at the generator. The refrigerant and solution are subsequently recombined in

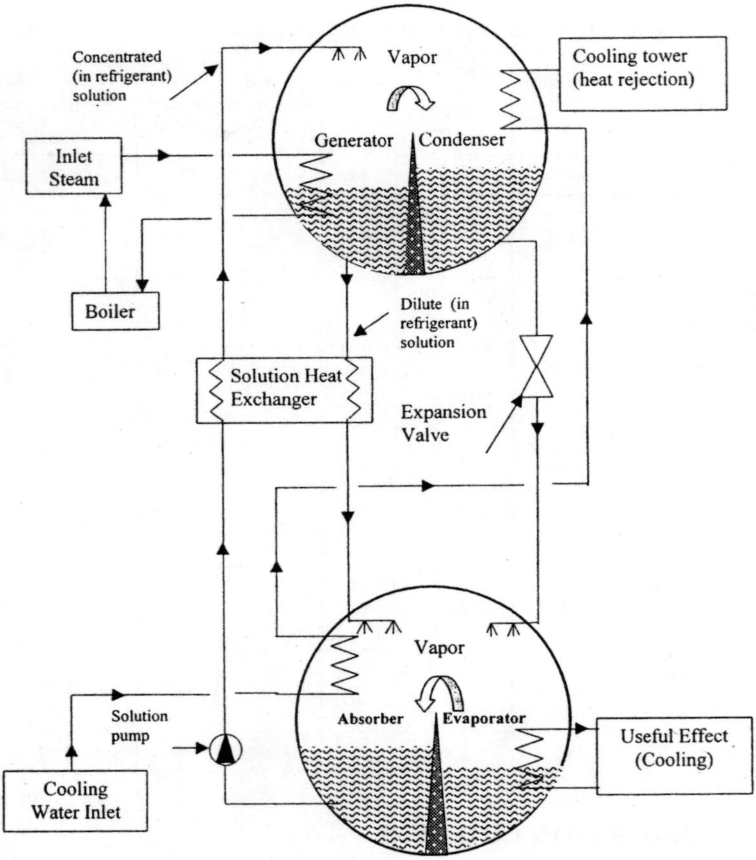

Figure 2.18: Schematic of an absorption chiller cycle. This particular illustration is for a single-stage steam-fired unit with a non-volatile solute as in the LiBr–water pair. The same schematic applies equally well to a hot water-fired device. Also, the heat pump mode involves extracting the useful effect as heating at the condenser/absorber, as opposed to the cooling mode where the useful effect is heat removal at the evaporator.

an exothermic process at the absorber. The absorber functions as a heat rejection unit (in addition to the condenser). Were the absorber to operate adiabatically, the solution temperature would increase, and after some time no vapor would be absorbed. Therefore heat rejection at the absorber is essential to cycle operation.

Before the liquid enters the generator, its pressure is elevated by a relatively low-power liquid pump (the pump's electrical power con-

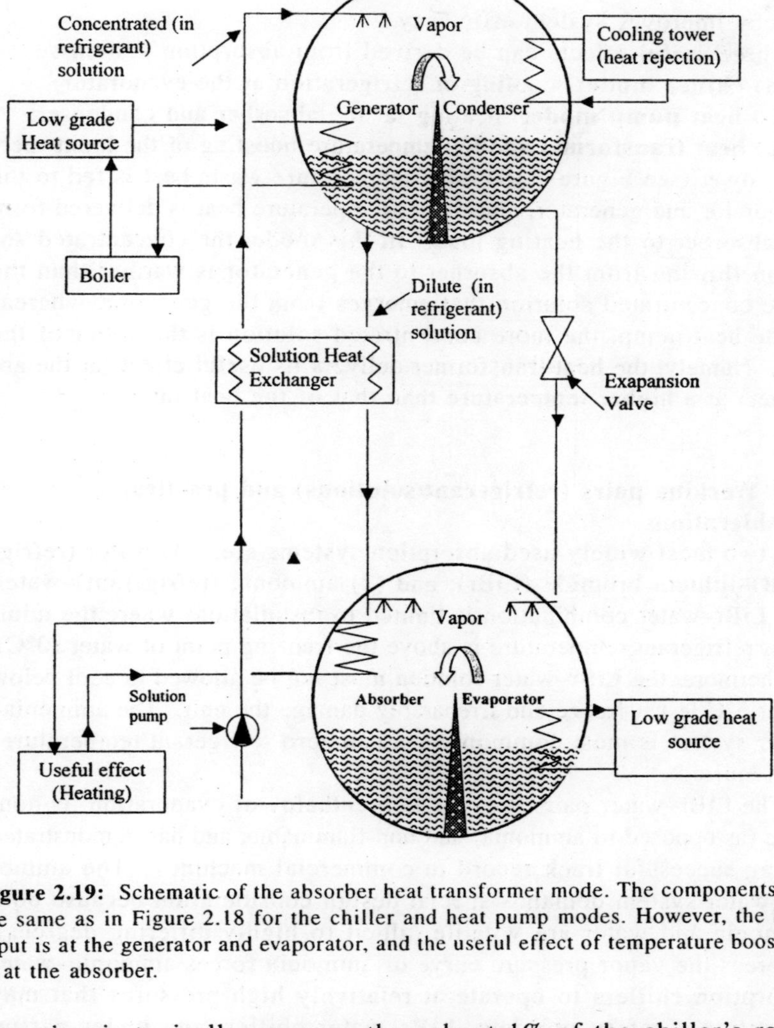

Figure 2.19: Schematic of the absorber heat transformer mode. The components are the same as in Figure 2.18 for the chiller and heat pump modes. However, the heat input is at the generator and evaporator, and the useful effect of temperature boosting is at the absorber.

sumption is typically no more than about 1% of the chiller's rated cooling capacity). In order to maintain a pressure difference between the generator and absorber, the solution is expanded (throttled) into the absorber.

A solution heat exchanger is introduced between the absorber and the generator. This additional heat exchanger is a heat recovery unit. It transfers heat *from* the warmer stream that exits the generator (diluted with respect to the refrigerant), *to* the colder stream the leaves the absorber (concentrated with respect to the refrigerant). This regenerative unit reduces the heat input requirement at the generator and thereby improves system efficiency.

Three useful effects can be derived from absorption systems:

(a) **chiller mode:** cooling or refrigeration at the evaporator;

(b) **heat pump mode**: heating at the absorber and condenser;

(c) **heat transformer mode**: temperature boosting of the input thermal power (see Figure 2.19). Low-temperature waste heat is fed to the evaporator and generator, and higher-temperature heat is delivered from the absorber to the heating load. In this mode, the concentrated solution flowing from the absorber to the generator is warmer than the more concentrated solution that emerges from the generator, whereas in the heat pump, the more concentrated solution is the hotter of the two. Namely, the heat transformer delivers its useful effect (at the absorber) at a higher temperature than that of the heat input.

C2. Working pairs (refrigerant solutions) and practical considerations

The two most widely used absorption systems are: (1) water (refrigerant)–lithium bromide (LiBr); and (2) ammonia (refrigerant)–water. The LiBr–water combination is limited to installations where the minimum refrigerant temperature is above the freezing point of water (0°C). Furthermore, the LiBr–water solution must not be allowed to cool below about 5°C lest it freeze and irreparably damage the unit. The ammonia–water system is most common when sub-zero refrigerant temperatures are required.

The LiBr–water pair enjoys a high enthalpy of evaporation, is non-toxic (as opposed to ammonia) and non-flammable, and has demonstrated a long successful track record in commercial machines. The ammonia–water system demands special design consideration because both ammonia and water are volatile (albeit to highly differing degrees). Whereas the vapor pressure curve of ammonia forces ammonia–water absorption chillers to operate at relatively high pressures that may constitute a safety problem, LiBr–water chillers run under partial vacuum.

Absorption devices have exhibited long lifetimes and excellent part-load behavior. The common working pairs of LiBr–water and ammonia–water are non-ozone-depleting and non-global warming chemicals, in contrast to standard refrigerants in mechanical chillers.

The major limitations of absorption machines, relative to their mechanical counterparts, are restricted temperature ranges and relatively high initial costs. The COPs of absorption machines are also markedly lower than those of mechanical chillers. But meaningful compari-

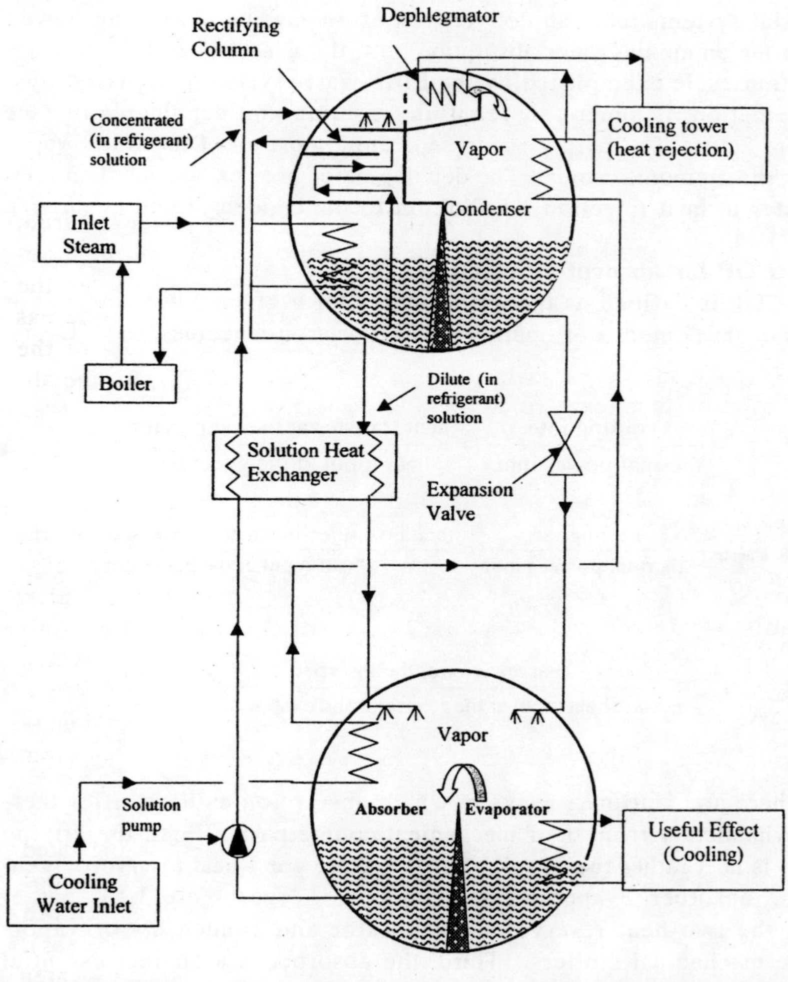

Figure 2.20: Schematic of an absorption chiller when both components of the binary mixture are volatile, as in the ammonia–water system. The illustration is for a single-stage system. Note the introduction of a rectifying column and a dephlegmator between the generator and the condenser, in order to achieve adequate separation of the binary mixture. The dephlegmator contributes to heat rejection.

sons among COPs require a close examination of the source of primary energy, in order to account properly for the inherent conversion losses from primary fuel to electricity in the operation of mechanical chillers.

Commercial absorption chillers range in size from small domestic units with cooling rates of the order of hundreds of watts, to large industrial systems that can deliver tens of megawatts of cooling power.

In the ammonia–water absorption unit, the generator of the basic absorption cycle exemplified by the LiBr–water system is replaced by a combination of generator, rectifying column and dephlegmator (see Figure 2.20). The purpose is to separate almost all of the water vapor from the ammonia vapor. The dephlegmator heat exchanger then contributes to heat rejection (in addition to the condenser and absorber).

C3. COP for absorption machines

The COP is defined as the useful effect for a given power input. For each of the 3 modes of operation of absorption machines, the COP is:

$$\mathrm{COP}_{\text{chiller}} = \frac{\text{cooling rate}}{\text{thermal power input}} = \frac{\text{heat transfer at the evaporator}}{\text{heat input at the generator}} \tag{2.10}$$

$$\mathrm{COP}_{\text{heat pump}} = \frac{\text{heating rate}}{\text{thermal power input}} = \frac{\text{total heat rejection at condenser and absorber}}{\text{heat input at the generator}} \tag{2.11}$$

$$\mathrm{COP}_{\text{heat transformer}} = \frac{\text{heat rejection at the absorber}}{\text{total heat input at the generator and evaporator}}. \tag{2.12}$$

There are 5 primary ways in which absorption chillers differ thermodynamically from their mechanical counterparts. First, the driving force is heat rather than work. Second, there are 4 heat reservoirs (generator, absorber, evaporator and condenser – see Figure 2.18), rather than the two heat reservoirs (evaporator and condenser) of vapor-cycle mechanical chillers. Third, the absorber is a distinct essential extra component. Fourth, when the total heat rejection is treated as constrained, one has an extra control variable in the division of the heat rejection between the absorber and condenser. And fifth, there is significant internal dissipation (entropy production) from the mass-transfer processes inherent to the absorption cycle and from regenerative heat exchangers when they are introduced.

The COP of absorption systems is ostensibly far lower than that of the corresponding mechanical devices. This apparent inferiority, however, stems from absorption devices processing thermal power directly – hence incurring entropy production in the conversion of thermal power to the useful effect. Equivalently, the COP of mechanical chillers is defined so that the entropy produced in generating the electrical power input is viewed as external, *i.e.*, not accounted for. Otherwise the COP for conversion of primary fuel into cooling power is comparable for mechanical and absorption systems, with the advantage to the mechanical machines (where central power generation can be carried out efficiently). The key advantage of absorption technology lies in the direct utilization of locally-available thermal sources, and in reduced environmental and pollution dangers.

The big advantage of absorption systems is that a liquid is compressed instead of a vapor. Hence compression work is negligible in absorption systems, whereas it is typically 20–50% of the rated cooling capacity in vapor-compression devices. The nominal virtues of heat-driven systems (as opposed to mechanical work-driven systems) are offset, however, by: (1) a markedly lower COP; (2) size and complexity; and (3) expense.

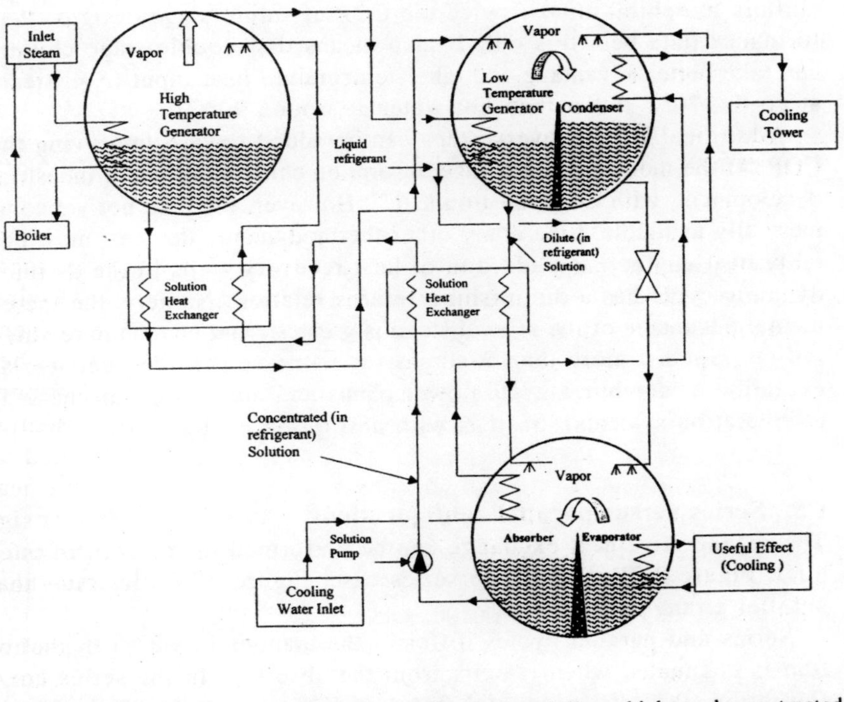

Figure 2.21: Schematic of a double-stage absorption chiller, which can be contrasted with the single-stage configuration drawn in Figure 2.18. The regenerative heat exchange is effected in a series (rather than a parallel) configuration.

C4. Heat regeneration and multi-stage configurations

Current commercial absorption chillers come in single and double-stage cycles (triple-stage cycles are under commercial development). Figure 2.18 shows a single-stage cycle, and for illustrative contrast, Figure 2.21 depicts a double-stage machine. The number of stages refers to the number of heat recovery units (generator heat exchangers) at different temperatures. Regenerative heat exchange improves COP by lessening the thermal bottleneck of finite-rate heat and mass exchange.

Single-stage absorption chillers use the thermal power input once and only once to generate heated vapor from the refrigerant–solution mixture. In distinction, *double*-stage chillers generate twice, in two separate generators. The higher-pressure refrigerant vapor generated in the first stage is condensed; the heat of condensation is exploited (partly recovered) to generate lower-pressure refrigerant vapor from the solution a second time.

The heat recovery or heat regeneration in the double-stage chiller generates almost twice the amount of refrigerant vapor as in the corresponding single-stage unit. Hence one would expect double-stage chillers to exhibit almost twice the COP of single-stage designs. Performance data bear this out. It also means that double-stage chillers can take better advantage of higher-temperature heat input (*e.g.*, steam at up to 170°C rather than hot water at around 90°C).

Additional heat recovery steps can be added toward improving the COP. At the moment, triple-stage absorption chillers are under industrial development, with COPs of around 1.7. However, they are not yet commercially available. Like many other thermodynamic devices, most notably heat engines, the addition of heat-recovery steps in the thermodynamic cycle has a diminishing returns relation. Namely, the incremental advantage of the second stage is greatest, and economic realities militate against more than 3 stages (as witnessed in the commercial evolution of combined-cycle power plants and multi-stage mechanical refrigeration systems). So it is with absorption cycles too.

C5. Series versus parallel configurations

The regenerative heat exchange can be performed in series or in parallel. Figure 2.21 showed the series case. Figure 2.22 illustrates the parallel arrangement.

Series and parallel cycles differ in the manner in which the solution is channeled when it exits from the absorber. In the series configuration, the refrigerant-rich solution is first pumped to the high-temperature generator, partially separated, and then directed to the low-temperature generator for further separation. In the parallel assembly,

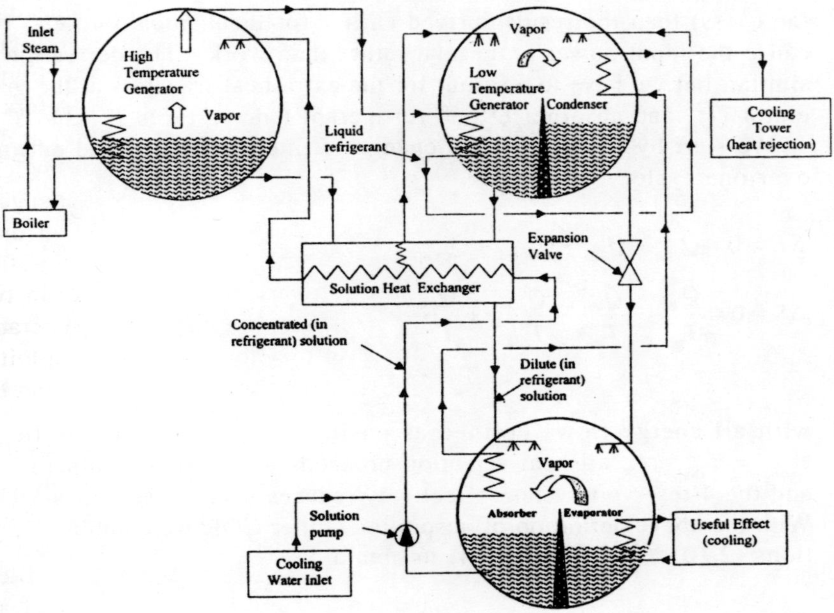

Figure 2.22: Schematic showing parallel versus series regenerative heat exchange in a double-stage absorption chiller. In contrast to Figure 2.21 for series regeneration, note that there is only one solution heat exchanger here.

the solution is split in the regenerator (the solution heat exchanger) and delivered to both the high and low-temperature generators. The two heated streams are later reunited at the regenerator.

An advantage of the parallel cycle is a higher COP and a lower risk of crystallization (for the LiBr–water pair) than the series cycle. The parallel cycle can accept lower pressure steam than the series cycle, which reduces the required input thermal power for the same cooling rate. The key disadvantage of the parallel cycle is royalty costs for its recently-granted patent.

We will return to a more thorough examination of regenerative absorption chillers in Section C of Chapter 12.

C6. Derivation of fundamental bounds for absorption COP

Next we proceed to the derivation of the fundamental limits to the thermodynamic performance of absorption machines. The reversible Carnot limit for the COP of absorption chillers must inherently be different (and less) than the result derived earlier for mechanical chillers, because the input power is thermal rather than work. The derivation is similar, but we have to account for the extra heat transfers at the generator Q_{gen} and absorber Q_{abs} at refrigerant temperatures T_{gen} and T_{abs}.

We start by calculating the change in internal energy and entropy over one cycle:

$$\Delta E = 0 = Q_{\text{gen}} - Q_{\text{abs}} - Q_{\text{cond}} + Q_{\text{evap}} \tag{2.13}$$

$$\Delta S = 0 = \frac{Q_{\text{gen}}}{T_{\text{gen}}} - \frac{Q_{\text{abs}}}{T_{\text{abs}}} - \frac{Q_{\text{cond}}}{T_{\text{cond}}} + \frac{Q_{\text{evap}}}{T_{\text{evap}}} \tag{2.14}$$

with all energy flows defined as positive. In the reversible limit, $T_{cond} = T_{abs}$, *i.e.*, all heat rejection proceeds at the same temperature, and the 4-reservoir system effectively reduces to a 3-reservoir system. With the above definition of absorption chiller COP, we combine Equations (2.10), (2.13) and (2.14) to obtain

$$\text{COP}^{\text{Carnot}}_{\text{absorption chiller}} = \frac{\dfrac{1}{T_{\text{abs}}} - \dfrac{1}{T_{\text{gen}}}}{\dfrac{1}{T_{\text{evap}}} - \dfrac{1}{T_{\text{abs}}}}. \tag{2.15}$$

For the absorption heat pump, with the useful effect being the total heat rejection, it follows that

$$\text{COP}^{\text{Carnot}}_{\text{absorption heat pump}} = \text{COP}^{\text{Carnot}}_{\text{absorption chiller}} + 1. \tag{2.16}$$

For the absorption heat transformer, the useful effect is the heat rejection at the absorber only, and the input is the total heat absorption at the generator and evaporator. The reversible limit again reduces a 4-reservoir system to a 3-reservoir system, but with a distinct condenser and absorber whereas the generator and evaporator now operate at the same temperature. It then follows that

$$COP^{Carnot}_{absorption\ heat\ transformer} = \frac{\frac{1}{T_{cond}} - \frac{1}{T_{evap}}}{\frac{1}{T_{cond}} - \frac{1}{T_{abs}}} \quad (2.17)$$

Tutorial 2.3

Table 2.4 shows the technical specifications for a steam-fired single-stage LiBr–water absorption chiller. The rated conditions are T^{in}_{evap} = 12.7°C, T^{in}_{cond} = 30.0°C and T^{in}_{gen} = 115.0°C. The chiller generates a cooling power of 3068 kW.

When we refer to a "dilute" or "concentrated" solution here, we mean dilute or concentrated *with respect to LiBr*. This is the convention adopted by chiller engineers in dealing with LiBr–water absorption machines. (Unfortunately, in relating to ammonia–water absorption machines, chiller engineers use "dilute" to indicate dilute with respect to the refrigerant.)

Table 2.5, taken from standard thermodynamic tables [ASHRAE 1998], provides the thermodynamic properties of the LiBr–water solution at the state points along the absorption cycle.

Figure 2.23 sketches the principal chiller components, heat flows and flow rates.

Based on these specifications and invoking simple mass balance at the generator,

Table 2.4: Technical specifications of the steam-fired single-stage LiBr–water absorption chiller.

variable	value
evaporator temperature	5°C
absorber equilibrium temperature	45°C
solution temperature in the absorber (and at the inlet to the heat exchanger)	45°C
dilute solution temperature after the solution heat exchanger (before entering the generator)	82°C
concentrated solution temperature before the solution heat exchanger (after exiting the generator)	104°C
concentrated solution temperature after the solution heat exchanger (before throttling into the absorber)	60.6°C
refrigerant (water) vapor temperature leaving the generator	98°C
refrigerant (water) liquid temperature leaving the condenser	45°C
mass fraction of LiBr in the solution leaving the generator	0.652
mass fraction of LiBr in the solution returning to the generator (from the absorber)	0.603

Table 2.5: Thermodynamic properties of the refrigerant (water) at key points along the absorption refrigeration cycle. (Pressure values are included to afford an appreciation of the partial vacuum under which the chiller operates.)

state point	pressure (kPa)	temp. (°C)	specific enthalpy (kJ kg^{-1})	specific entropy (kJ kg^{-1} K^{-1})	LiBr mass fraction, X
1. inside the generator	9.81	104	261.39	0.5262	0.652
2. solution heat exchanger outlet (concentrated solution)	0.89	60.6	185.2	0.31182	0.652
3. inside the absorber	0.87	45	126.10	0.2438	–
4. absorber outlet	0.87	45	126.10	0.2438	0.603
5. solution heat exchanger outlet (dilute solution)	9.81	82	196.64	0.4536	0.603
6. generator outlet	9.81	98	2683.05	8.4617	0.0
7. condenser outlet	9.59	45	188.41	0.6385	0.0
8. evaporator outlet					
vapor	0.87	5	2509.71	9.0235	0.0
liquid	0.87	5	188.41	0.67807	0.0

determine:

(1) the ratio of solution mass flow rate at the absorber to refrigerant mass flow rate, commonly called the circulation flow rate ratio CR;

(2) the chiller's COP; and

(3) the mass flow rate of saturated steam that must be supplied to the generator.

Solution: We begin by considering the mass balance for the refrigerant flows into and out of the generator (at steady state):

(mass flow rate of refrigerant entering from the concentrated solution)
– (mass flow rate of refrigerant leaving via the dilute solution)
= (mass flow rate of refrigerant leaving as vapor from the generator) . (2.18)

To work with convenient dimensionless variables, we: (a) divide each of the mass flow rates by the refrigerant mass flow rate; and (b) express the relative amount of LiBr in the solution by its mass fraction X. Equation (2.18) can then be expressed as

$$(1 - X_{conc})\,\text{CR} - (1 - X_{dilute})\,(\text{CR} - 1) = 1 \qquad (2.19)$$

where the subscripts "conc" and "dilute" refer to the concentrated and dilute solutions, respectively. With the given values of $X_{conc} = 0.603$ and $X_{dilute} = 0.652$, we solve Equation (2.19) for CR, to obtain CR = 13.31.

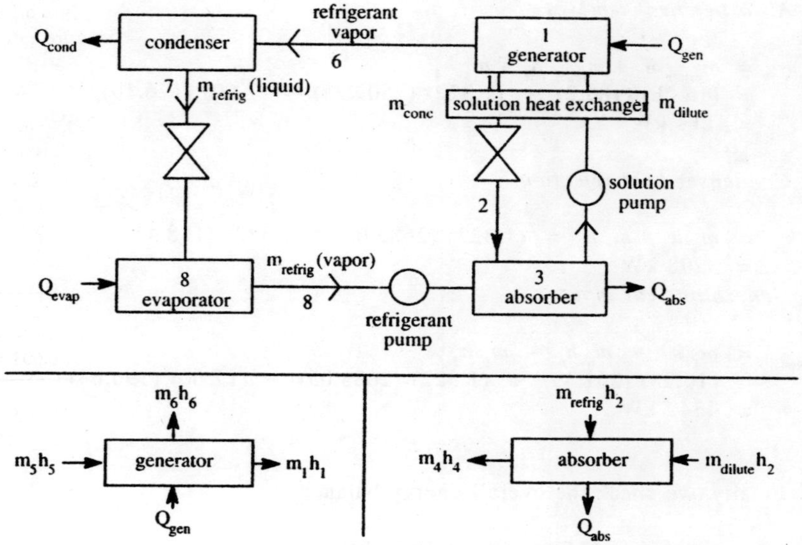

Figure 2.23: Schematic of the absorption cycle, highlighting the components, heat flows and flow rates used in solving the problem. Subscripts on variables reflect the state point numbering of Table 2.5.

The refrigerant mass flow rate m_{refrig} is obtained from the prescribed cooling power and the specific enthalpy values in Table 2.4:

$$m_{refrig} = \frac{Q_{evap}}{h_8 - h_7} = \frac{3068}{2509.71 - 188.41} = 1.322 \text{ kg s}^{-1}.$$

Then the mass flow rate of the concentrated and dilute solutions, m_{conc} and m_{dilute}, respectively, are

$$m_{conc} = m_{refrig} \text{ CR} = (1.322)\ (13.31) = 17.60 \text{ kg s}^{-1}$$

$$m_{dilute} = m_{conc} - m_{refrig} = 17.60 - 1.322 = 16.28 \text{ kg s}^{-1}.$$

Now we are set to calculate the principal heat flows in the cycle. We have been given the evaporator heat extraction, Q_{evap} = 3068 kW. The absorber, condenser and generator heat flows are calculated as follows (with the numerical subscripts referring to the state points listed in Table 2.5).

1) *Absorber heat removal*

$$\begin{aligned} Q_{abs} &= m_{dilute}\, h_2 + m_{refrig}\, h_8 - m_4\, h_4 \\ &= (16.28)\,(185.2) + (1.322)\,(2509.71) - (17.60)\,(126.10) \\ &= 4114 \text{ kW}. \end{aligned}$$

2) *Condenser heat rejection*

$$\begin{aligned} Q_{cond} &= m_6\, h_6 - m_7\, h_7 = (1.322)\,(2683.05) - (1.322)\,(188.41) \\ &= 3298 \text{ kW}. \end{aligned}$$

3) *Generator heat input*

$$\begin{aligned} Q_{gen} &= m_1\, h_1 + m_6\, h_6 - m_5\, h_5 \\ &= (16.28)\,(261.39) + (1.322)\,(2683.05) - (17.60)\,(196.64) \\ &= 4342 \text{ kW}. \end{aligned}$$

Finally, we check the overall energy balance:

$$\begin{aligned} Q_{in} &= Q_{gen} + Q_{evap} = 4342 + 3068 = 7410 \text{ kW} \\ Q_{out} &= Q_{cond} + Q_{abs} = 3298 + 4114 = 7412 \text{ kW}. \end{aligned}$$

According to the First Law, Q_{in} and Q_{out} should be equal (the very small difference here is due to round-off error).

D. THERMOACOUSTIC CHILLER

In thermoacoustic refrigeration, high-intensity sound waves are used instead of compressors to set up a standing wave in a closed resonator tube filled with inert gases, and in which a stack of plates is inserted with heat exchangers at its ends [Swift 1988; Garrett & Hofler 1992] (see Figure 2.24). The gas is compressed by the acoustic standing wave, warms up, and transfers heat to the stack plates. The temperature difference that develops along the stack plates is called the thermoacoustic effect. A heat exchanger rejects part of this heat, and the remaining cooled gas is used to chill the load via the other heat exchanger. The process is cyclic.

The basic but involved physics and thermodynamics underlying thermoacoustic processes are already well understood [Wetzel & Herman 1997]. The two predominant irreversibilities are viscous dissipation in the working fluid, and finite-rate heat transfer at the heat exchangers.

The most notable use of the thermoacoustic refrigerator to date has been as a cryocooler in satellites [Garrett & Hofler 1992], where using low input power and having large temperature spans (100–200 K) are critical (in contrast, for example, to commercial mechanical chillers

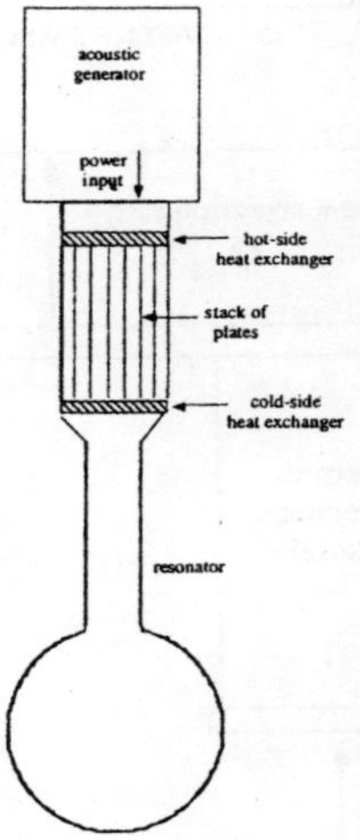

Figure 2.24: Schematic of a thermoacoustic chiller.

which have far higher input power and far smaller temperature spans). We'll return to the thermoacoustic chiller in Chapter 10 to examine how its performance data compare with the universal aspects of the chiller models that will be developed in the ensuing chapters.

E. THERMOELECTRIC CHILLER

When an electrical current *I* is passed through two dissimilar thermoelectric materials (denoted by A and B in Figure 2.25, usually metal or semiconductor alloys), one heats up while the other grows colder. Referred to as the Peltier effect, it forms the basis for thermoelectric refrigeration [Ioffe 1957; Goldsmid 1960].

A principal virtue of thermoelectric chillers is that they are solid state devices with no moving parts and no fluids. They accept DC power input, and can be temperature controlled with great precision. The niche applications for thermoelectric devices are miniaturized cooling loads, unlike conventional mechanical chillers. They are commonly used in military, aerospace, consumer product and medical instrument applications, among others. The dimensions of just the thermoelectric module itself are typically 2.5 × 2.5 × 0.5 cm, and some commercial units are as small as 0.4 × 0.4 × 0.2 cm. A complete commercial thermoelectric chiller package may occupy only around 300–500 cm^3 of space.

Typically, commercial thermoelectrics comprise semiconductors, most commonly bismuth telluride. The semiconductor material is doped to

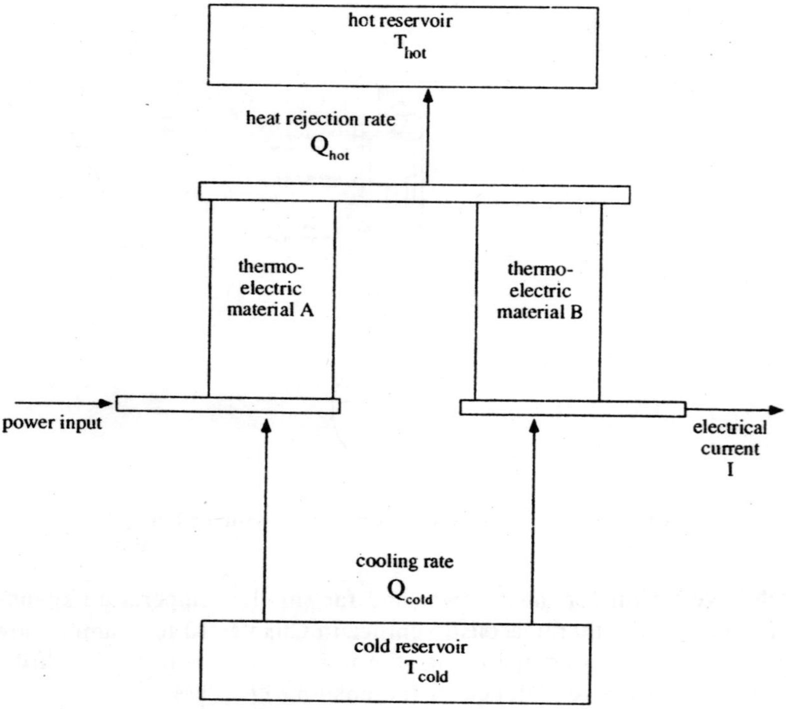

Figure 2.25: Schematic of a thermoelectric chiller

produce an excess of electrons in one element (n-type), and a dearth of electrons in the other element (p-type). Electrical power input drives electrons through the device.

At the cold end, electrons absorb heat as they move from a low energy level in the p-type semiconductor to a higher energy level in the n-type element. At the hot side, electrons pass from a high energy level in the n-type element to a lower energy level in the p-type material, and heat is rejected to a reservoir.

Figure 2.25 shows the simplest two-element configuration (referred to as a single couple). Commercial thermoelectric chillers designed for large thermal lifts ($T_{hot} - T_{cold}$), *e.g.*, larger than 40°C, are often built with two or more couples (4 or more elements), that are connected electrically in series and thermally in parallel. This is somewhat analogous to the construction of double- or multi-stage mechanical and absorption chillers.

The thermoelectric couple can be characterized by three material-dependent properties, its: (1) differential thermoelectric power coefficient α (sometimes referred to as the differential Seebeck coefficient) in units of V K^{-1}; (2) total electrical resistance R_{el}; and (3) total thermal conductance K. These properties are approximated here as independent of temperature.

Practical thermoelectric devices typically suffer negligible losses due to finite-rate heat transfer at the junctions (*i.e.*, negligible relative to the other irreversibilities). The heat flows and the division of electrical resistive losses between the two reservoirs are arrived at by solving the governing heat conduction equation with internal heat generation [Ioffe 1957; Goldsmid 1960]. The net cooling rate, Q_{cold}, in transferring heat from the cold reservoir at temperature T_{cold} to the hot reservoir at temperature T_{hot} is given by

$$Q_{cold} = \alpha I T_{cold} - K(T_{hot} - T_{cold}) - \frac{I^2 R_{el}}{2}. \tag{2.20}$$

The heat rejection to the hot reservoir Q_{hot} is

$$Q_{hot} = \alpha I T_{hot} - K(T_{hot} - T_{cold}) + \frac{I^2 R_{el}}{2}. \tag{2.21}$$

Hence from the First Law, the electrical power input P_{in} is

$$P_{in} = Q_{hot} - Q_{cold} = \alpha I (T_{hot} - T_{cold}) + I^2 R_{el}. \tag{2.22}$$

$$COP = \frac{Q_{cold}}{P_{in}}. \tag{2.23}$$

The points of maximum cooling rate and maximum COP are readily calculated from Equations (2.20)–(2.23).

The sources of irreversibility in the thermoelectric refrigerator are electrical resistance and heat leak. The heat leak militates against slow operation, *i.e.*, low electrical current, and electrical resistance mitigates against fast operation, *i.e.*, high current. In Chapters 10, 13 and 14, we'll come back to the thermoelectric chiller to probe how its thermodynamic performance can be compared to that of other chiller classes.

Chapter 3

STANDARDS, MEASUREMENTS AND EXPERIMENTAL TEST FACILITIES FOR CHILLERS AND HEAT PUMPS

"Scientific apparatus offers a window to knowledge, but as they grow more elaborate, scientists spend ever more time washing the windows."

Isaac Asimov

A. INTRODUCTION

The rest of this book is predicated on a familiarity with chiller and heat pump standards, and with the experimental measurements taken on these devices. For the reader who may not be well acquainted with the associated laboratory procedures, we devote this chapter to a review of standards and measurement techniques, along with the presentation of a relatively new testing method that overcomes much of the expense and complication of current standard test facilities.

Chiller standards also stipulate measurement tolerances for temperatures, flow rates, and power inputs. We'll review these stringency requirements and translate them into typical experimental uncertainties for the determination of cooling rates and COPs.

B. THE BASICS OF STANDARDS

B1. Wherefore standards?

Chillers and heat pumps need to be tested following certain standards prior to their release into the market. Standards provide a valuable basis for device evaluation, irrespective of whether the machines are reciprocating, centrifugal, absorption, adsorption or otherwise, and independent of whether the useful effect is cooling or heating.

Why are standards necessary? Basically, the reason is to level the playing field. Chiller manufacturers tend to present product information or performance data in a manner which favors their products. For example, a chiller or heat pump exhibits better performance if T^{in}_{cond} and T^{in}_{evap} are set close to ambient temperature. Unfortunately, the performance rating of a chiller or heat pump operating at such conditions will rarely be useful. The unsuspecting customer, for whom the actual load conditions are invariably remote from those of the ambient state, could easily be misled. Commercial standards offer a fair platform where all chiller and heat pump manufacturers can present customers with the performance of their products at realistic conditions.

B2. Types of standards

There are two kinds of standards that have been developed for chillers and heat pumps: (1) commercial standards, which are issued by professional engineering societies such as the American Society for Heating and Refrigeration Engineers (ASHRAE), the American Refrigeration Institute (ARI, an industrial body that sets industrial standards for air conditioning and refrigeration equipment), and the International Institute of Refrigeration (IIR), among others; and (2) statutory standards, issued by governmental organizations. The commercial standards issued by ASHRAE, ARI and IIR are similar for a given type of chiller or heat pump, with the ARI standards tending to be the most thorough. Accordingly, only ARI standards are cited below.

Commercial standards are: (1) non-binding for the introduction of new products; (2) formulated for the professional societies that issue them by chiller manufacturers, practicing engineers, installers, contractors, users and researchers; and (3) subjected to periodic review and amendment as the state of the art in the industry advances. *Statutory* standards are prescribed by national standards institutions which may demand a stipulated performance level to be met prior to the machine's approval or certification for sale. Statutory standards often relate more to the mechanical integrity and robustness of the machine than to its thermal performance. In this chapter, the standard and rating conditions from commercial standards only are discussed since they deal primarily with *thermodynamic* performance.

B3. What constitutes commercial standards?

The test methods outlined in most commercial standards treat a chiller as a blackbox that can only be probed externally, *i.e.*, non-intrusively. Only the coolant temperatures, coolant flow rates and power input supplied to the test unit are measured and recorded during a test (see Figure 3.1). The key variables to be controlled are: the coolant inlet temperatures T^{in}_{cond} and T^{in}_{evap} the coolant mass flow rates supplied to the heat exchangers m_{cond} and m_{evap}; and the electrical power consumption of the compressor and fluid pumps. These variables suffice because the two quantities to be determined are the machine's:

(*a*) useful effect = $m\ C^{*}\Delta T$, where ΔT is the temperature difference across the heat exchanger (the evaporator for chillers and the condenser for heat pumps), and C is the coolant's specific heat for the temperature range of interest; and

(*b*) $$\text{COP} = \frac{\text{useful effect}}{\text{power input}}.$$

(For absorption machines, the *thermal* power input measured is that of the heat source, be it gas-fired, steam-fired or hot-water fired.)

The schematic of Figure 3.1 for mechanical chillers also suffices for mechanical heat pumps provided the positions of the evaporator and condenser are interchanged. Figure 3.2 is the corresponding test rig schematic for absorption machines.

Commercial standards provide either the *standard* ratings or the *application* ratings. Rated conditions are selected to be useful to a pro-

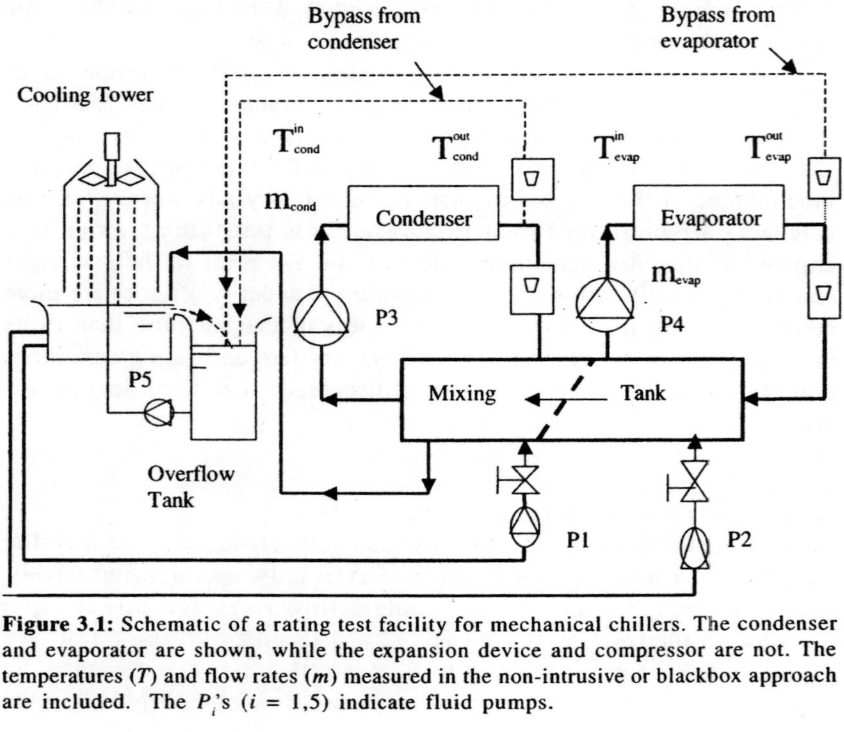

Figure 3.1: Schematic of a rating test facility for mechanical chillers. The condenser and evaporator are shown, while the expansion device and compressor are not. The temperatures (T) and flow rates (m) measured in the non-intrusive or blackbox approach are included. The P_i's (i = 1,5) indicate fluid pumps.

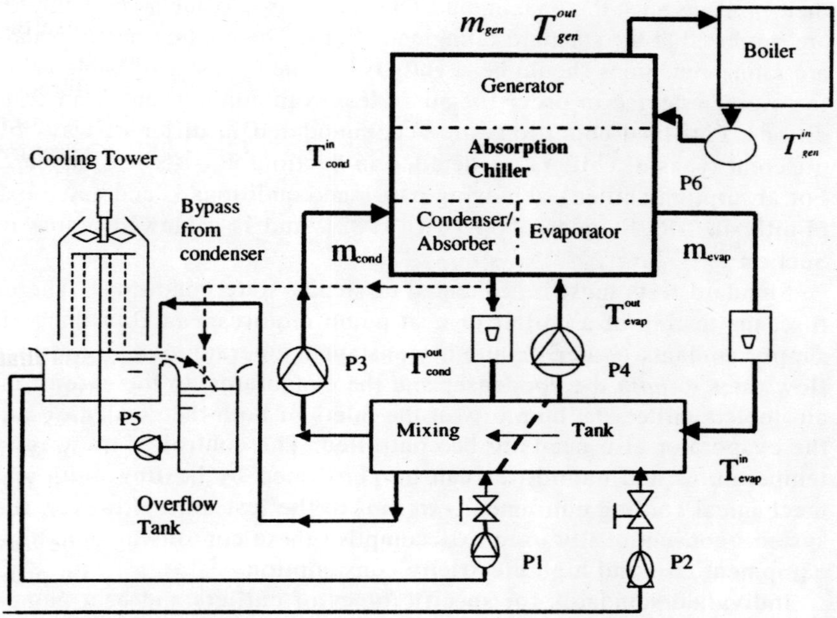

Figure 3.2: Schematic of a rating test facility for absorption chillers. The generator, evaporator and heat rejection unit (condenser/absorber) are shown, while the expansion device is not. The temperatures (*T*) and flow rates (*m*) measured in the non-intrusive or blackbox approach are included. The P_i's (i = 1,5) indicate fluid pumps.

spective user in energy performance comparisons among different products. A typical example of standard rating conditions for a water-cooled chiller includes T^{in}_{cond}= 29.4°C and T^{in}_{evap} = 12.2°C. In contrast, *application* ratings provide the performance of a product when operating at other than standard rating conditions that arise from major variations in ambient conditions, load conditions and/or changes in the application. An example where application ratings prove necessary is chiller operation in semi-arid or desert regions where the ambient dry-bulb temperature during part of the year can reach as high as 45°C. Clearly, the performance of a chiller subjected to such ambient extremes would not be fairly reflected by standard ratings. In the ARI standards, the application ratings for semi-arid and desert regions are occasionally termed Standard Rating B, although these ratings are rarely quoted by manufacturers.

Standard rating conditions for chillers reflect coolant variable values for typical air-conditioning (space cooling) loads. The acceptance of the standard gives chiller manufacturers a strong incentive to design their products such that maximum COPs and near-maximum cooling rates are produced at the standard rating conditions. The cooling rate at standard rating conditions should be slightly below the largest achievable value for a given device in order to satisfy less common extreme load conditions. Part-load conditions are accommodated in different ways by different types of chillers, as detailed in Sections B4-B6 of Chapter 2. For absorption chillers, achieving part-load conditions is addressed explicitly in ARI Standard 560 [ARI 1982], and is reviewed below in Section F.

Standard tests must be executed at steady-state conditions. Therefore, the testing of a chiller or heat pump requires a facility that can supply coolants (water or air) at constant temperatures *and* constant flow rates to *both* the condenser *and* the evaporator. In the case of an air-cooled chiller, the humidity at the inlets of both the condenser and the evaporator also needs to be controlled. The control of these inlet temperatures and humidities can be performed by heating units and mechanical cooling equipment *external* to the test unit. However, the systems conventionally used to accomplish these controls involve high equipment cost and high electricity consumption.

Individual standards for specific types of chillers and heat pumps will be reviewed in Sections E-G. In Section H, we'll present a novel alternative procedure whereby the maintenance of the inlet variables of the coolant is achieved by: (1) mixing the coolant leaving the evaporator and the condenser of the unit being tested; and (2) a special mixing strategy for the coolant supplied from the cooling tower or from ambient. It is this alternative test procedure and facility that was employed to generate some of the performance data reported in Chapters 4 and 6, for predictive and diagnostic tools that demand the relatively low experimental uncertainties commensurate with ARI chiller standards. This level of stringency in measurement accuracy may be fulfilled by the data points actually measured and reported by manufacturers, but may not be satisfied by nominal data points produced by the extrapolation and/or interpolation of those measurements. We will return to this issue in Section B3 of Chapter 6 when we consider the impact of measurement precision on the determination of chiller parameters regressed with fundamental thermodynamic modeling.

C. DESIGNING AN EXPERIMENTAL TEST FACILITY

We can outline the test conditions for conducting standard and application rating tests for chillers and heat pumps without regard to how one designs and assembles the test facility. In planning a test facility, one may opt for one of two design classes, depending on one's needs:

(1) A facility with a capability only for external (non-intrusive) measurements of coolant states: the blackbox approach (as in Figures 3.1 and 3.2). The internal states of the refrigerant are not measured and chiller performance is inferred from steady-state performance at a specified set of coolant flow rates and inlet/outlet temperatures. As such, the accurate determination of chiller COP is acutely sensitive to the experimental uncertainties of the coolant temperatures and flow rates. This type of experimental facility is preferable when chillers are tested according to commercial standards, because the internal (refrigerant) flows cannot be tempered, so the results obtained non-intrusively are similar to those of the nominal installed state of a chiller.

(2) A facility which includes internal measurements of the thermodynamic states of the *refrigerant*, as well as external measurements of the coolant states. This more complex facility is necessary in the research laboratory where the additional measurements of refrigerant states permit the computation of essential chiller characteristics which could not be derived from the blackbox approach. Chiller designers *are* usually interested in the extra information that can be gleaned, such as the heat exchanger thermal conductances, the process average temperatures, and the internal entropy production of the principal components. This additional information affords a diagnostic capability that stems from establishing the link between the coolant states and system losses, and can be used to regress for key component variables. These are central issues to which we'll repeatedly be returning throughout the book.

The design, optimization and diagnostic capabilities that derive from this second probing approach, as well as descriptions of the associated test facilities, are documented for the principal chiller types in the chapters that follow. Verification of the predictive and diagnostic capabilities of the thermodynamic models presented in these chapters necessitates data from both external and internal measurements, even if the eventual implementation of those models can suffice with the blackbox approach.

D. MEASUREMENT ACCURACY, INSTRUMENTATION AND EXPERIMENTAL UNCERTAINTY

In most commercial standards, the maximum uncertainty (*i.e.*, the acceptable tolerance) for a temperature measurement, and hence for maintaining a fixed temperature, is ±0.28°C. This implies that a Class A type of sensor is necessary in all temperature readings (although it is not mentioned explicitly in the standards themselves).

The time interval to determine a steady-state period is normally taken

Table 3.1: Summary of rating conditions and operating limits for reciprocating, centrifugal and absorption chillers. Details of the standards cited are reviewed in Section E.

	reciprocating chillers ARI Standard 590 [ARI 1986c]	centrifugal chillers ARI Standard 550 [ARI 1986a]	absorption chillers ARI Standard 560 [ARI 1986b]
condenser: temperatures	$T_{cond}^{in} = 29.4°C$ $T_{cond}^{out} = 35.0°C$	$T_{cond}^{in} = 29.4°C$ $T_{cond}^{out} = (NS)$	$T_{cond}^{in} = 29.4°C$ $T_{cond}^{out} = (NS)$
volumetric coolant flow rate per kW of cooling power	(NS)	0.054 l s^{-1} kW^{-1}	0.069 l s^{-1} kW^{-1}
evaporator: temperatures	$T_{evap}^{in} = 12.2°C$ $T_{evap}^{out} = 6.7°C$	$T_{evap}^{in} = (NS)$ $T_{evap}^{out} = 6.7°C$	$T_{evap}^{in} = (NS)$ $T_{evap}^{out} = 6.7°C$
volumetric coolant flow rate per kW of cooling power	(NS)	0.043 l s^{-1} kW^{-1}	0.043 l s^{-1} kW^{-1}
	acceptable tolerances and fluctuations at steady state		
temperature	± 0.28°C		
flow rate	± 5% of specified value		
pressure	(NA)	(NA)	for steam-fired units: ± 1.4 kPa in inlet steam pressure
achieving steady-state	Steady-state period of at least 30 minutes, preceded by 20 minutes of steady-state operation at the same tolerances		

(NS = not stated, NA = not applicable)

Table 3.2: A list of typical measurement accuracies for the variables measured in chiller and heat pump test facilities. Refrigerant variables are included to cover the more comprehensive type of test facility. δm, $\delta(\Delta T)$, δQ_{evap}, and δP_{in} denote the uncertainties in the coolant mass flow rate, temperature difference across the heat exchanger, cooling rate and input power, respectively. Coolant specific heat C is assumed to be numerically exact.

variable	estimated error	remarks
temperature	± 0.21°C comprised of ± 0.15°C for the sensor, times √2 to account for the random error in the temperature measurement	Class A Resistance-Thermal-Device(RTD) temperature sensors are required, equivalent to the specification outlined in instrumentation standards DIN/IEC 751 or British Standard 60751. *e.g.* for an RTD, the minimum conformance is a 4-wire device. The ΔT across the heat exchanger is about 3-5°C for most designs. The uncertainty in temperature measurement for Tutorials 3.1 and 3.2 is hence about 6%.
pressure	0.5% of full span	An accurate pressure gauge can be used if pressure recording is not required.
refrigerant mass flow rate	1–2% of full span	Refrigerant flow rate can be measured in either the liquid or gaseous phase. Liquid phase measurements are more stable, particularly for steady-state experiments.
coolant mass flow rate	0.25% of full span	The coolant is either water or air.
solution *refrigerant* concentration (for absorption machines)	0.5% of full span	Usually reported to 3 significant figures.
cooling rate and COP (mechanical chillers)	$Q_{evap}=C(m\pm\delta m)(\Delta T\pm\delta(\Delta T))$ $\delta Q_{evap}=mC\Delta T\sqrt{\left[\frac{\delta m}{m}\right]^2+\left[\frac{\delta(\Delta T)}{\Delta T}\right]^2}$ $\mathrm{COP}=\frac{Q_{evap}\pm\delta Q_{evap}}{P_{in}\pm\delta P_{in}}=\frac{Q_{evap}}{P_{in}}\times\sqrt{\left[\frac{\delta Q_{evap}}{Q_{evap}}\right]^2+\left[\frac{\delta P_{in}}{P_{in}}\right]^2}$	Total uncertainty is estimated at around 7%. The contribution from electrical power measurement is negligible.
cooling rate and COP (absorption chillers)	Same as for mechanical chillers (immediately above), but with P_{in} replaced by Q_{gen}	Acceptable uncertainty values are about 5–7%.

to be at least 30 minutes, preceded by an additional period of 20 minutes at different conditions but with the same temperature tolerance level.

Table 3.1 summarizes the standard rating conditions and measurement tolerances for reciprocating, centrifugal and absorption chillers.

The basic instrumentation required in a simple facility for rating mechanical chillers must include measurements of flow rate, temperature and electrical power input. For absorption chillers, measurements of pressure and of the rate of thermal power input are also required. The total uncertainty for the COP can be estimated from the type of instrumentation used, with Table 3.2 listing typical error bands. Sample temperature–time traces from actual standard chiller and heat pump tests are shown in Figures 3.3 and 3.4 in Tutorials 3.1 and 3.2 below.

The differential temperature measurement across the heat exchangers tends to be the major source of error, because the heat exchangers usually have low differential readings of ΔT = 3-5°C. Despite the use of class A sensors, the uncertainty in this reading typically amounts to about 6%.

Flow rate measurements can commonly be made to around 1% accuracy. Errors from measuring input electrical power are usually negligible. Hence a simple test facility with basic instrumentation should provide a COP determination with an uncertainty level of about 7%. By employing expensive matched-pair temperature sensors for the en-

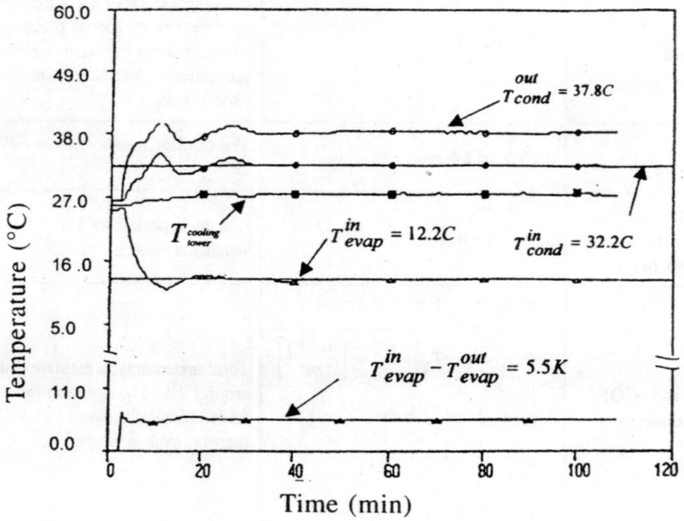

Figure 3.3: Temperature–time trace during an application rating test of a reciprocating chiller.

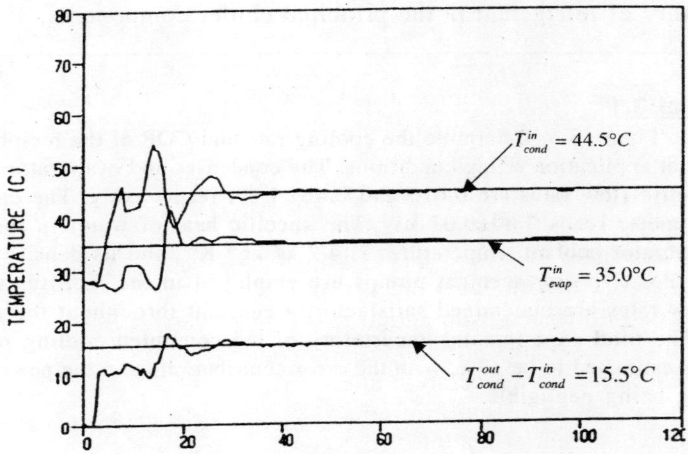

Figure 3.4: Temperature–time trace during the standard rating test of a reciprocating heat pump.

ergy flow computation, one can noticeably reduce the total uncertainty in the COP.

No measurement can be viewed as more accurate than its experimental uncertainty. For example, if, after proper error analysis, the COP of a heat pump is measured to be 3.10±0.30, then clearly any alteration in the heat pump that results in its COP changing by less than 0.30 cannot be accepted as statistically significant. The uncertainty in the determination of any variable is determined from the combination of systematic and random errors, in accordance with standard error analyses of experimental data (see, for example, ASHRAE Standard 41.5-75 [ASHRAE 1975]). The total uncertainty comprises contributions from each of the individual measurements of temperature, pressure, flow rate and power input, plus the assumed values for assorted material constants such as density, specific heat and thermodynamic state variables. Table 3.2 offers a list of typical accuracies for the variables measured in chiller and heat pump test rigs. These constitute the primary contributions to the total uncertainty in determining the useful effect (cooling or heating rate) and the COP.

When transient (as opposed to steady-state) chiller performance is desired, the response time of each sensor must be evaluated separately. The rule of thumb is that the sensor should have a response time an order of magnitude faster than the intended signal being measured. Furthermore, the comprehensive test facility that can monitor refrigerant variables must include gravimetric measurement devices for tracking the inventory of refrigerant in the principal chiller components.

Tutorial 3.1

Refer to Figure 3.3. Determine the cooling rate and COP of the reciprocating chiller at application rating conditions. The condenser and evaporator coolant volumetric flow rates are 0.616 and 0.461 l s^{-1}, respectively. The electrical power meter reads 3.80±0.01 kW. The specific heat of water in the range of evaporator coolant temperatures is 4.2 kJ kg^{-1} K^{-1}, and its density is 1.00 kg l^{-1}. Positive displacement pumps are employed in the facility, and water flow rates are maintained satisfactorily constant throughout the test period. The total experimental uncertainty of the computed cooling rate and COP is estimated to be ±7%, with the error contribution from the power meter reading being negligible.

Solution: First, from inspection of Figure 3.3, we confirm that, at the rating conditions, the steady-state interval is at least 30 minutes, in accordance with the requirements of ARI Standard 590. Next, from the data, we have the elementary calculations:

useful effect of cooling at the evaporator $= (\rho V)_{evap}\ C(T^{in}_{evap} - T^{out}_{evap})$
$= 1.00 \cdot 0.461 \cdot 4.2 \cdot 5.5 = 10.65$ kW

experimental uncertainty in cooling rate $= 0.07 \cdot 10.65 = 0.75$ kW

electrical input $= 3.8$ kW

COP $= \dfrac{10.65}{3.8} = 2.80$

experimental uncertainty in COP $= 0.07 \cdot 2.80 = 0.20$

Tutorial 3.2

Based on the application rating test shown in Figure 3.4, determine the COP of the heat pump when hot water is produced at 60°C at the condenser outlet. The evaporator is supplied with water at 35°C. As per the specification of the manufacturer, water volumetric flow rates through the condenser and evaporator are 0.33 and 0.37 l s^{-1}, respectively. The electrical power consumption is 5.72±0.01 kW. The specific heat of water in the range of the condenser coolant temperature is 4.2 kJ kg^{-1} K^{-1} and its density is 0.984 kg l^{-1}. The experimental uncertainty of the COP is ±7%. Calculate the useful effect and COP.

Solution: After checking the preliminaries as in Tutorial 3.1, we have:

useful effect of heating at
the condenser $C(T^{out}_{cond} - T^{in}_{cond}) = (\rho V)_{cond}\ C(T^{out}_{cond} - T^{in}_{cond})$

$= 0.984 \cdot 0.33 \cdot 4.2 \cdot 15.5 = 21.16$ kW

experimental uncertainty
in heating rate $= 0.07 \cdot 21.16 = 1.48$ kW

electrical power input $= 5.72$ kW

COP $= \frac{21.16}{5.72} = 3.70$

experimental uncertainty
in COP $= 0.07 \cdot 3.70 = 0.26$.

E. STANDARD FOR WATER-COOLED MECHANICAL CHILLERS

The ARI Standard 590 [ARI 1986c] prescribes the test conditions for the standard rating test of water-cooled *reciprocating* chillers. The main control variables for a standard rating test are:

Condenser water:	$T^{in}_{cond} = 29.4\,°C$	entering
	$T^{out}_{cond} = 35.9\,°C$	leaving
Evaporator water:	$T^{in}_{evap} = 12.2°C$	entering
	$T^{out}_{evap} = 6.7°C$	leaving.

The ARI Standard 550 [ARI 1986a] sets the test conditions for the standard rating test of *centrifugal* water-cooled chillers, for which the coolant variables are:

- $T^{out}_{evap} = 6.7°C$
- $T^{in}_{cond} = 29.4\,°C$
- chilled water volumetric flow rate per kW of cooling power $= 0.043$ l s^{-1} kW^{-1}
- condensed water volumetric flow rate per kW of cooling power $= 0.054$ l s^{-1} kW^{-1}.

Typical measurement tolerances for these and other chiller standards are summarized in Table 3.1. There are distinct differences in the requirements for conducting a rating test on reciprocating versus centrifugal machines. For a reciprocating chiller, the temperature difference between the inlet and outlet of the heat exchangers is specified, and the coolant flows in the evaporator and the condenser are adjusted to satisfy the stipulated temperatures.

In contrast, for a centrifugal unit, only the temperatures of the coolant leaving the evaporator and entering the condenser of the test unit are specified. The coolant temperatures entering the evaporator and leaving the condenser can assume any convenient values, provided the volumetric flow rates per kW of cooling rate are as specified in the test standard.

The reason for adjusting the water flow rates per kW of cooling power, rather than limiting the size of the heat exchangers (and hence constraining the temperature differences across the heat exchangers) is to minimize the pumping power invested in distributing coolant throughout the plant. Centrifugal chillers are usually designed for relatively large cooling loads: roughly in the range 500–10,000 kW. Coolant pipelines in such plants tend to be long in stretching between the air-handling units or cooling towers and the heat exchangers of the chiller. By comparison, reciprocating chillers are designed for smaller cooling loads, typically less than 350 kW. In such small systems, constraining heat exchanger area to optimal values becomes the dominant issue rather than operational costs.

F. ABSORPTION CHILLER STANDARD

The ARI Standard 560 [ARI 1986b] was developed for absorption chillers, with conditions for the evaporator and condenser that are similar to those of ARI Standard 550 for centrifugal chillers (refer to Table 3.1). Figure 3.2 is a schematic of a sample test rig for absorption machines. Since the absorber is normally cooled by coolant from the cooling tower (*i.e.*, from ambient conditions), the water flow rate through the condenser can be increased by as much as 25%, *i.e.*, typically as high as $0.068\ \mathrm{l\ s^{-1}\ kW^{-1}}$. This increase is independent of whether the absorber is cooled in series or in parallel with the condenser.

For a water-fired absorption chiller, the following measured variables are specified: the flow rate of the heat source; the temperatures of the entering and leaving coolant, and the pressure drop. For a steam or gas-fired chiller, the input flow rate and pressure (and steam quality) are specified along with the pressure drop across the pressure regulator.

Manufacturers also specify a "fouling factor" allowance: a quantity that refers to the anticipated increase in the heat exchanger's thermal resistance, that commonly occurs over the course of normal operating conditions. A typical value of the fouling factor allowance is $8.8 \times 10^{-5}\ \mathrm{m^2\ K\ W^{-1}}$.

In the ARI Standard 560, chiller part-load performance is determined on the basis that:

- $T_{evap}^{out} = 6.7°C$
- chilled water flow rate remains as per full load
- T_{cond}^{in} varies linearly with load from the temperature at full load down to 15.6°C
- coolant flow rates for the condenser and absorber remain constant as per full load.

In some climates, the low local ambient wet-bulb temperature can pose a difficulty in maintaining T_{cond}^{in} during part-load tests. When part-load conditions force T_{cond}^{in} to fall below 15.6°C, a bypass line across the condenser water piping can be used to raise this temperature.

Tutorial 3.3

A hot water-fired single-stage LiBr–water absorption chiller has a nominal rated capacity of 7.0 kW (2 Rton) at the following coolant temperatures and mass flow rates:

$$T_{gen}^{in} = 80.0°C \qquad T_{gen}^{out} = 78.2°C \qquad m_{gen} = 0.889\ kg\ s^{-1}$$

$$T_{cond}^{in} = 29.5°C \qquad m_{cond} = 0.463\ kg\ s^{-1}$$

$$T_{evap}^{in} = 10.0°C \qquad m_{evap} = 0.333\ kg\ s^{-1}.$$

These conditions represent part-load behavior (often called an application

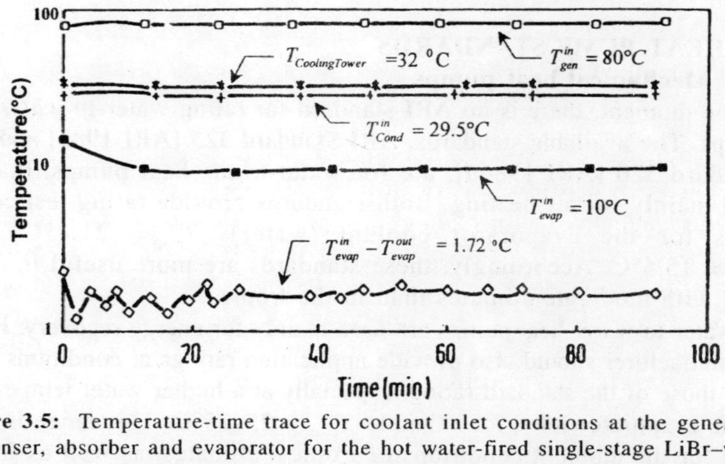

Figure 3.5: Temperature-time trace for coolant inlet conditions at the generator, condenser, absorber and evaporator for the hot water-fired single-stage LiBr–water absorption chiller being tested. Note the logarithmic ordinate scale, in order to discriminate visually among all the temperatures in the plot.

rating) that is well below the nominal rating. The solution specific heat is C = 4.2 kJ kg^{-1} K^{-1}.

The steady-state temperature–time trace of coolant water entering the generator, condenser, absorber and evaporator of the chiller are displayed in Figure 3.5.

For these rating conditions, determine the chiller's cooling rate and COP.

Solution: First, we check that the conditions of ARI Standard 560 are satisfied: coolant (water) inlet temperatures must be constant for a period of 30 minutes with temperature fluctuations of less than ±0.2 K. In addition, the steady-state period must be preceded by a period of 20 minutes at the same tolerances. From Figure 3.5, we confirm that these stipulations are indeed respected.

Second, we calculate the heat input to the generator:

$$Q_{gen} = (mC)_{gen}\,(T^{in}_{gen} - T^{out}_{gen}) = (0.889 \cdot 4.2)\,(80.0 - 78.2) = 6.72 \text{ kW.}$$

For the cooling rate at the evaporator, we can read the value of $T^{in}_{evap} - T^{out}_{evap}$ = 1.7°C off the graph (so T^{out}_{evap} = 8.3°C), and obtain

$$Q_{evap} = (mC)_{evap}\,(T^{in}_{evap} - T^{in}_{evap}) = (0.333 \cdot 4.2)\,(10.0 - 8.3) = 2.38 \text{ kW.}$$

$$\text{COP} = \frac{Q_{evap}}{Q_{gen}} = \frac{2.38}{6.72} = 0.354.$$

The experimental uncertainty for COP is estimated at ±5%, so the application rating COP is 0.354±0.018.

G. HEAT PUMP STANDARDS

G1. Mechanical heat pumps

At the moment, there is no ARI standard for rating water-to-*water* heat pumps. The available standards, ARI Standard 325 [ARI 1985] and ARI Standard 320 [ARI 1986d], are for water-to-*air* heat pumps, that are used mainly for air heating. Both standards provide rating test conditions for the evaporator coolant (water): T^{in}_{evap} = 21.1°C and T^{out}_{evap} = 15.5°C. Accordingly, these standards are more useful in countries with moderate climates than in the tropics.

Water-to-*water* heat pumps are used mainly for energy recovery. Hence a manufacturer should also provide application ratings at conditions other than those of the standard rating, especially at a higher water temperature at the evaporator inlet, for example T^{in}_{evap} = 25–35°C. This range is common for the ambient conditions of tropical climates, as well as for the conditions of industrial waste heat reclamation.

Common commercial heat pumps can supply hot water up to 60°C without an appreciable degradation of the properties of the refrigerant. It is inadvisable to rate heat pumps beyond a condenser inlet temperature of 60°C because the refrigerant can readily decompose after many cycles of operation. Although there is no specific standard for higher water temperatures, the hot water temperature of 60°C, which is the standard temperature for hot water leaving a heat pump's desuperheater, has been adopted by ARI Standard 470 [ARI 1987]. ARI Standard 470 relates to the performance of the desuperheating section of the condenser heat exchanger in a mechanical heat pump or chiller.

G2. Absorption heat pumps

Absorption heat pumps follow the rating requirements of ARI Standard 560. In a heat pump or heat transformer, the useful effect is heating at the condenser and/or absorber, while the coolant from ambient conditions (via the cooling tower or ground water) is piped to the heat pump's evaporator.

H. AN ALTERNATIVE TEST PROCEDURE AND MIXING STRATEGY

H1. Why bother with alternative test rig designs?

In this section, we describe a test facility which not only can provide the stringent requirements of constant coolant temperatures and flow rates to the evaporator and condenser, but also is economical to operate in terms of equipment costs and electricity consumption. This facility

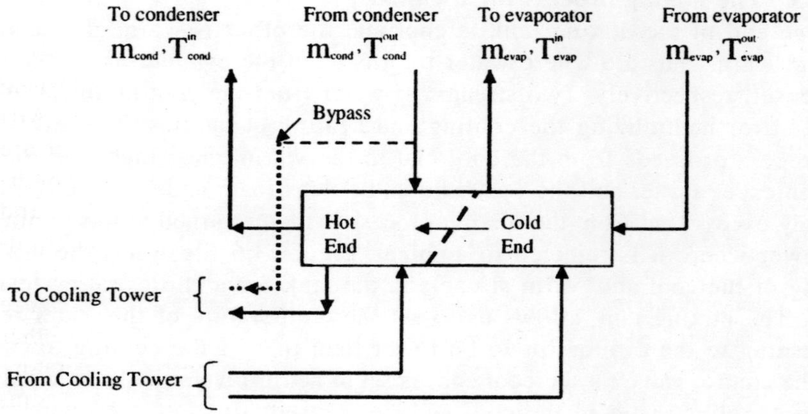

Figure 3.6: Schematic illustration of the mixing process for a chiller rating test.

dispenses with the need for the conventional storage of large volumes of hot and cold water at the required rating conditions prior to the commencement of a test run. A second advantage is that our alternative test rig allows for faster turnaround times for a given rating test. We'll delineate the operation of a water-to-water cooled chiller, but the principles involved can easily be extended to other cooling media, as well as to other types of chillers and heat pumps. Furthermore, the facility described here is not merely a proposal; it has been built, tested and verified as an accurate, inexpensive alternative to conventional chiller test installations [Bong *et al* 1989, Bong *et al* 1990].

H2. The basic idea for simplifying the procedure

A mixing tank is used to produce the required temperatures at the entrance to the evaporator and condenser of the test unit (chiller or heat pump, as in Figures 3.6 and 3.7). The essence of this procedure is to neutralize the heating and cooling capacities that are produced by the condenser and evaporator of the test unit. The mixing of these streams is performed in a manner such that only the desired temperature levels are returned to the evaporator and the condenser. Based on temperature measurements at pre-selected locations, the mixing processes are executed by computer-controlled modulating valves using a simple Proportional-Integral-Differential control strategy. For effective control of *extreme* conditions, as required by application ratings, the mixing tank is designed with bypass pipelines so that part of the chilled water and/or warm water can circumvent the mixing process.

H3. The mixing process for a chiller

One end of the mixing tank is cool and the other is warm. The cool and warm ends are where water returns from the evaporator and condenser, respectively. Two streams of water from the cooling tower are used for neutralizing the cooling and heating capacities. The mixing process proceeds from the cool end to the warm end which, for convenience, are termed the upstream and downstream ends, respectively. Any excess heat from the mixing process is then returned to the cooling tower where it is rejected to ambient. Figure 3.6 illustrates the mixing of the cool and warm streams in the tank for a chiller rating test.

The mixing tank allows us to set the temperature of the water returning to the evaporator to be lower than that of the cooling tower. The control valve (at the cool end) is set to act in reverse so as to supply *less* cooling water to the cool end. In contradistinction, the temperature of the water returning to the condenser can be raised by the control valve at the warm end, where the valve control strategy can act in either the forward or reverse direction based on the feedback from the measured temperatures. A bypass of the condenser water to the cooling tower may be necessary if T^{in}_{cond} is lower than that of the temperature of the cooling tower. Figure 3.3 is a temperature–time trace from a chiller rating test performed with the alternative test rig in accordance with ARI Standard 590.

H4. Mixing process for a heat pump

Rating tests for a heat pump require higher supply temperatures of coolant water to both the evaporator and the condenser of the test unit. We retain the same mixing tank as for the chiller rig, but proceed from the warm end to the cool end (instead of vice versa). Typical parameters for a rating test with a heat pump are illustrated in Figure 3.7.

With this new mixing strategy, the condenser can be supplied with water at a temperature above that obtainable from the cooling tower. Similarly, the evaporator can be supplied with water at a temperature above, as well as below, that derived from the cooling tower. The control valve at the warm end operates in the forward mode when the temperature of water returning to the condenser is higher than that of the cooling tower. In rating tests where T^{in}_{evap} is set *above* that of the cooling tower, bypass water from the condenser outlet is reduced to a minimum. Conversely, should T^{in}_{evap} be set *lower* than that of the cooling tower water, the bypass from the condenser outlet is allowed to open, and the cor-

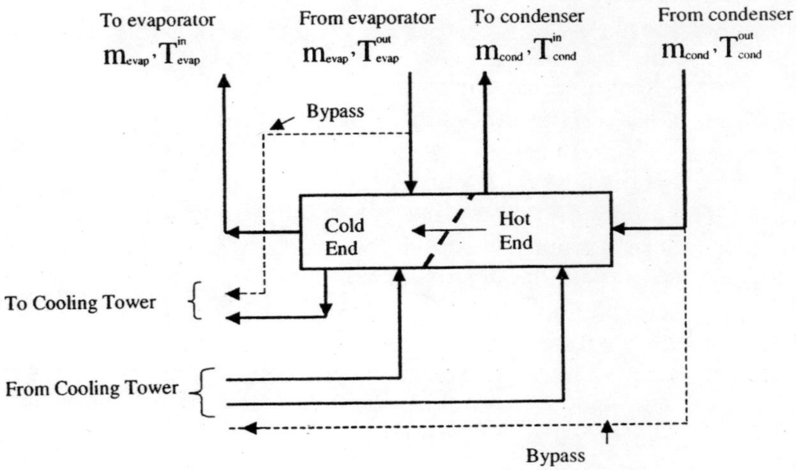

Figure 3.7: Schematic illustration of the mixing process for a heat pump rating test.

responding control valve at the cool end operates in the reverse mode. Figure 3.4 is a temperature–time trace during a rating run in the alternative test facility.

Chapter 4

ENTROPY PRODUCTION, PROCESS AVERAGE TEMPERATURE AND CHILLER PERFORMANCE: TRANSLATING IRREVERSIBILITIES INTO MEASURABLE VARIABLES

"A child of five would understand this. Send somebody to fetch a child of five."
- Groucho Marx

A. ENTROPY PRODUCTION

The simplest and most basic limit to chiller thermodynamic performance is the upper bound on COP that follows from the Second Law (Equation (2.5)). Between the reversible limit

$$COP_{Carnot} = \frac{T_{cold}}{T_{hot} - T_{cold}} \tag{4.1}$$

and the actual COP of real chillers lies a sea of irreversibilities. These dissipative mechanisms turn out to be so great (at least with technologies developed to date) that in reality the COPs of commercial chillers rarely exceed half of the reversible limit of Equation (4.1). Small reciprocating chillers realize far smaller fractions of the Carnot COP. As a quantitative example, we offer in Figure 4.1 performance data from a relatively efficient commercial reciprocating chiller. The ratio of actual to Carnot COP varies from 0.07 to 0.21, depending on operating conditions (*i.e.*, depending on coolant temperatures and cooling rate). From experimental measurements of chiller performance in manufacturer catalogs, from several of the references cited in this book and from actual chiller data reported directly here, the reader can confirm that these low fractions of COP_{Carnot} are typical of commercial chillers.

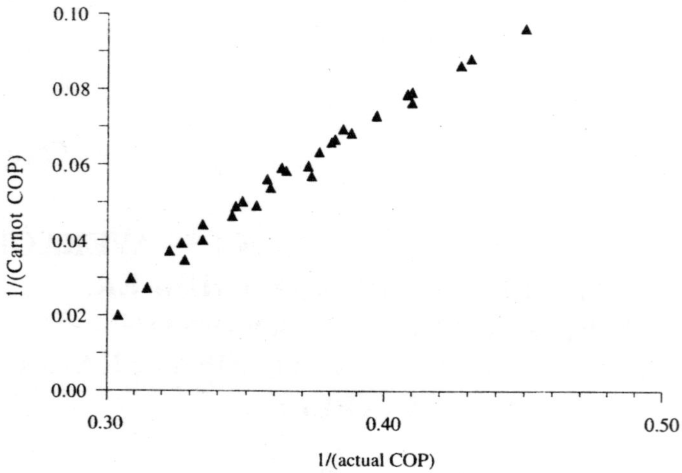

Figure 4.1: A plot of 1/(Carnot COP) against 1/(actual measured COP) for a relatively efficient commercial reciprocating chiller, for the full range of chiller operating conditions: cooling rate = 8.9–13.4 kW; T^{in}_{cond} = 23.81–35.05°C; and T^{in}_{evap} = 7.98-18.02°C. The reversible limit is based on coolant inlet temperatures: $COP_{Carnot} = \frac{T^{in}_{evap}}{T^{in}_{cond} - T^{in}_{evap}}$. Data are from [Chua *et al* 1996].

Entropy production is another way of saying irreversibility or dissipation. It is a clearly-defined thermodynamic variable that can be determined with measured quantities. We will divide irreversibilities into 3 general classes: external, internal and heat leaks. External losses derive from finite-rate heat transfer between the refrigerant and the coolants, *i.e.*, from the bottleneck associated with the chiller's thermal communication with its reservoirs. Internal dissipation refers to the entropy production that does not stem from the chiller's interaction with its environments. Heat leaks are the parasitic heat transfers between the refrigerant and its surroundings.

In mechanical chillers, internal dissipation is dominated by frictional losses in the compressor, with modest contributions from the throttler, de-superheating in the condenser, and small pressure drops in the heat exchangers. These loss mechanisms also contribute to internal losses in absorption chillers. The internal irreversibilities that are unique to absorption chillers stem from chemical potential drops (losses in the chemical potentials of the refrigerant and solution as a consequence of finite-rate mass transfer and dissipative mixing effects), and all losses in regenerative heat exchangers being internal because their heat exchange involves no thermal communication with the coolants (reservoirs).

In this chapter, the impact of external losses on COP will be accounted for implicitly since COP will be expressed in terms of refrigerant (rather than reservoir) temperatures. We postpone explicit treatment of the effect of coolant temperatures on chiller COP until Chapter 5. We also focus here upon internal losses. As we'll see in Sections E and F, one can de-couple the impact of the 3 classes of irreversibility and express COP as the sum of independent contributions from each.

How do we translate internal entropy production in a chiller into the associated increase in the input power required to achieve a given cooling rate? In this chapter, we'll review the rudiments of this "translation", and in the process establish an approach that simplifies, clarifies and unifies the thermodynamic analysis of chillers. The key is what we will refer to as Process Average Temperature (PAT). As we'll see shortly, the PAT is the conversion factor for calculating the incremental input power requirement (equivalently, lost work) from the corresponding entropy production. The PAT can also be interpreted as the temperature of the refrigerant were the process to be viewed as an equivalent isothermal process, for purposes of calculating entropy changes. As such, the PAT is not a measured temperature; rather it is the proper reference temperature for evaluating dissipative losses.

B. EXAMPLE FOR MECHANICAL CHILLERS

Let's start with a non-reactive single-component system, as one has in mechanical chillers. All our analyses are restricted to steady-state processes (*i.e.*, no transient effects). One way of expressing the Second Law for a given process is

$$\Delta S_{\text{int}} = \int_{\text{in}}^{\text{out}} dS - \int_{\text{in}}^{\text{out}} \frac{dH}{T} \geq 0 \tag{4.2}$$

where: (*a*) ΔS_{int} is the internal entropy production; (*b*) the integrations are from inlet to outlet conditions; (*c*) the integral over dS is the change in the entropy of the refrigerant in the process; and (*d*) the integral over dH/T is the entropy transfer associated with external heat transfer with H denoting the enthalpy and T the temperature of the refrigerant. Values of S and H in Equation (4.2) can be found in standard tables of the refrigerant's thermodynamic properties [Mayhew & Rogers 1971, ASHRAE 1998], with a knowledge of the initial and final states of the refrigerant for the particular process.

A second independent means lies at our disposal for determining ΔS_{int}. For example, consider losses due to fluid friction - and hence a pressure drop – in a chiller component. Knowing how the refrigerant's volume V and temperature T depend on pressure p, one can calculate

$$\Delta S_{int} = -\int_{in}^{out} \frac{V dp}{T} \geq 0 \tag{4.3}$$

where dp is the (negative) pressure drop. Equations (4.2) and (4.3) must yield the same result.

A tutorial example that illustrates precisely how ΔS_{int} is calculated from standard experimental measurements is offered below in Section D.

C. EXAMPLE FOR ABSORPTION CHILLERS

The same exercise pertains to the more complex situation in absorption chillers where there are multiple streams and hence multiple phases. Each of a total of j inlet streams and k outlet streams can be viewed as an autonomous phase, denoted here by the subscript β. Each phase is comprised of more than one component.

For each component in a given process, we can again calculate ΔS_{int} in two equivalent, independent ways. The expression for the multi-phase system corresponding to Equation (4.2) is

$$\Delta S_{int} = \sum_{\beta=1}^{j} (m_\beta s_\beta)_{in} - \sum_{\beta=1}^{k} (m_\beta s_\beta)_{out} + \int_{in}^{out} \frac{dH}{T} \tag{4.4}$$

where m = mass flow rate, s = specific entropy and

$$dH = d\left[\sum_{\beta=1}^{j} (m_\beta h_\beta)_{in} - \sum_{\beta=1}^{k} (m_\beta h_\beta)_{out}\right] \tag{4.5}$$

with h denoting specific enthalpy. The analog of Equation (4.3) for a given component is

$$\Delta S_{\text{int}} = -\sum_{\text{all streams }\beta} \int_{\text{in}}^{\text{out}} \left[\frac{m_\beta v_\beta \mathrm{d}p}{T_\beta} + \frac{\mu_\beta \mathrm{d}m_\beta}{T_\beta} \right] \tag{4.6}$$

where v = specific volume and μ = chemical potential. Mass balance imposes the constraints

$$\int_{\text{in}}^{\text{out}} \mathrm{d}m_\beta = m_\beta \tag{4.7}$$

and

$$\sum_{\text{allstreams }\beta} m_\beta = \text{constant}. \tag{4.8}$$

Again, Equations (4.4) and (4.6) must yield the same result. A tutorial example that covers the calculation of ΔS_{int} for absorption systems is included in the following section.

D. PROCESS AVERAGE TEMPERATURE

If the individual contributions to entropy production can be identified experimentally, why the need for defining and working with a PAT? If one's sole aim is to ascertain COP at particular operating conditions, then in fact PAT is an unnecessary variable. However, once one is intent upon performing chiller diagnostics, or predicting chiller performance under different operating conditions, or evaluating COP improvements that would derive from diminishing a given source of irreversibility, then the need for an accurate PAT becomes essential.

The PAT is computed from the properly weighted piecewise compilation of measured temperatures along non-isothermal paths. When the PAT is multiplied by the entropy production in a given process, the lost potential work is obtained

$$\text{PAT} = \frac{\text{lost potential work}}{\text{entropy production}}. \tag{4.9}$$

There would be no problem in evaluating PAT in a known non-isothermal process if the temporal and spatial distributions of the thermodynamic properties in the process path were known. In practical situations, however, we are expected to relate to the thermodynamic system as an effective blackbox which can be probed from the outside only, *i.e*, for which only non-intrusive measurements at the inlets and outlets are realistic.

In terms of measurable thermodynamic variables, the PAT is given by

$$\mathrm{PAT} = \frac{\int_{\mathrm{in}}^{\mathrm{out}} \mathrm{d}H}{\int_{\mathrm{in}}^{\mathrm{out}} \frac{\mathrm{d}H}{T}}. \tag{4.10}$$

For mechanical chillers, recalling Equation (4.2), we can also express Equation (4.10) as

$$\mathrm{PAT} = \frac{\int_{\mathrm{in}}^{\mathrm{out}} \mathrm{d}H}{\int_{\mathrm{in}}^{\mathrm{out}} \mathrm{d}S - \Delta S_{\mathrm{int}}}. \tag{4.11}$$

The corresponding relation for absorption chillers (recall Equations (4.4)-(4.5)) is

$$\mathrm{PAT} = \frac{\int_{\mathrm{in}}^{\mathrm{out}} \mathrm{d}\left[\sum_{\beta=1}^{j} (m_\beta h_\beta)_{\mathrm{in}} - \sum_{\beta=1}^{k} (m_\beta h_\beta)_{\mathrm{out}}\right]}{\sum_{\beta=1}^{j} (m_\beta s_\beta)_{\mathrm{in}} - \sum_{\beta=1}^{k} (m_\beta s_\beta)_{\mathrm{out}} - \Delta S_{\mathrm{int}}}. \tag{4.12}$$

Many authors have adopted a PAT that is the ratio of the enthalpy change to the entropy change in the process - what is often called the entropic-average temperature T^{s}_{av}:

$$T^{s}_{av} = \frac{\int_{in}^{out} dH}{\int_{in}^{out} dS}. \tag{4.13}$$

A problem arises because Equation (4.13) ignores the contribution of internal dissipation.

All the terms that enter the right-hand side of Equation (4.10) can readily be computed from the thermodynamic properties of the refrigerant if the local pressures and temperatures at the inlets and outlets are known. For example, for purely sensible heat exchange (and no internal dissipation) from initial temperature T_i to final temperature T_f, Equation (4.10) reduces to the familiar mean-logarithm expression

$$\text{PAT} = \frac{T_f - T_i}{\ln \frac{T_f}{T_i}} \tag{4.14}$$

provided the fluid specific heat is constant. When the fluid specific heat has a non-negligible dependence on pressure or temperature, the process path should be divided into distinct processes where pressure and temperature measurements are available.

In a chiller's evaporator, the entropy transfer during heat exchange Q_{evap} can be expressed as $\Delta S_{evap} = Q_{evap}/T_{evap}$, where T_{evap} is the process-average refrigerant temperature. When the pressure drop in the heat exchanger is negligible (and fluid specific heat is approximately constant over the temperatures traversed), T_{evap} is given by Equation (4.14) for known inlet and outlet temperatures. Had the same heat transfer with the same entropy transfer been effected isothermally, the effective constant refrigerant temperature would be the PAT, T_{evap}. The same argument applies, of course, to heat exchange at the condenser and, for absorption chillers, at the generator and the absorber.

The PAT for internal dissipation in a mechanical chiller can be understood on simple physical grounds. The heat generated by internal dissipation creates a heightened heat rejection requirement at the condenser. Therefore the process-average value at the condenser T_{cond} is the appropriate PAT. For the absorption chiller, the PAT for internal dissipation will be comprised of the proper entropy-weighted contributions of all the reservoirs that participate directly in heat rejection: the condenser, absorber and generator. The exact formula will be derived below in Section F.

Although the PAT is not a measured temperature, it must lie within the experimentally measured range of temperatures for the process. This simple physical consistency check has not always been applied in chiller studies. In Chapter 12, in which we take a close look at internal dissipation in chiller heat exchangers, we'll show that previous studies have adopted incorrect PATs that fall outside the range of measured temperatures and give rise to subtle errors in chiller diagnostics. The entropic-average T^{s}_{av} of Equation (4.13) can lie outside the permitted bounds, whereas the correct PAT of Equation (4.10) must, by its very formulation, satisfy this physical consistency check.

At this stage, we offer separate tutorials for a reciprocating and an absorption chiller. Approximate methods are adopted to estimate the PATs in the reciprocating chiller, so that no additional information, such as results from simulation studies, is needed. However, it turns out that such approximations prove inadequate for the absorption system. The reason lies in the strong dependence of the solution's thermodynamic properties on its concentration, with that concentration changing noticeably along the heat exchangers. Accordingly, in the absorption chiller tutorial, we used a computer simulation to calculate accurate PAT values which in turn are used in estimating the individual contributions to the internal entropy production.

Tutorial 4.1

Table 4.1 tabulates experimental readings of pressure and temperature, along with standard thermodynamic properties, for the refrigerant freon R12 in a rated 10 kW reciprocating chiller, at each of the chiller's components. The state points enumerated in Table 4.1 are indicated on the *T–S* diagram for the cycle in Figure 4.2. The refrigerant mass flow rate is 0.084 kg s^{-1}.

(1) Compute the process average temperatures for the condenser and evaporator.

(2) Compute the rate of internal entropy production in the condenser, evaporator, compressor and throttler, and, from their sum, the rate of total internal entropy production in the chiller. Assume that compression and expansion are executed adiabatically.

Solution:

(a) We calculate the PATs for the heat exchangers in an approximate piecewise fashion from sums over all relevant cycle branches (itemized from 1 to n). The PAT for heat exchanger j is then expressed as

Table 4.1: Experimental measurements and standard thermodynamic properties for the refrigerant R12 in the nominal 10 kW reciprocating chiller

state point	pressure (bar)	temperature (K) ± 0.15 K	specific entropy s (kJ kg^{-1} K^{-1})	specific enthalpy h (kJ kg^{-1})
1. compressor discharge	10.345±0.034	360.73	1.6423±0.0007	402.7±0.2
2. condenser inlet (vapor)	10.31±0.138	359.49	1.638±0.001	401.0±0.2
3. saturated vapor phase, based on pressure measurement at point 2	–	–	1.545±0.003	369.8±0.6
4. saturated liquid phase, based on pressure measurement at point 5	–	–	1.142±0.015	242.5±4.7
5. condenser outlet	10.172±0.207	301.71	1.095±0.002	227.6±0.7
6. throttle inlet	–	302.45	1.098±0.002	228.4±0.7
7. evaporator inlet	3.414±0.035	276.01	1.10±0.01	228.4±0.7
8. saturated vapor phase, based on pressure measurement at point 9	–	–	1.560±0.002	351.1±0.4
9. evaporator outlet	2.793±0.034	279.08	1.580±0.001	356.8±0.1
10. compressor suction	2.655±0.034	280.12	1.587±0.001	357.0±0.1

$$\mathrm{PAT}_j = \frac{\int_{\text{inlet}}^{\text{outlet}} \mathrm{d}H}{\sum_{i=1}^{n} \frac{\int_{\text{branch } i} \mathrm{d}H}{T_i}}.$$

Referring to Table 4.1 and following the pertinent branches, we can estimate

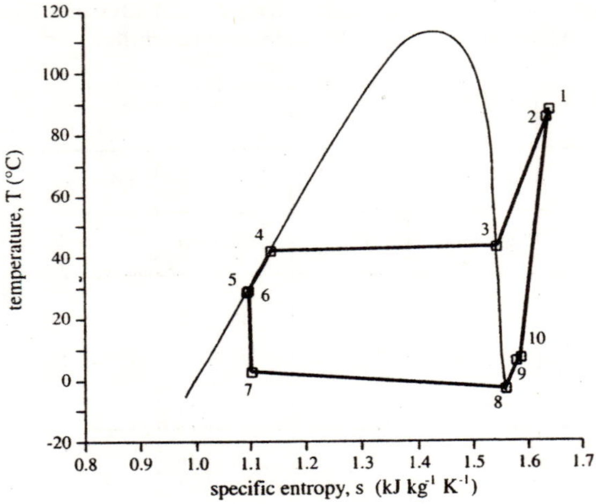

Figure 4.2: The *T–S* diagram for the refrigeration cycle, including the coexistence curve for the refrigerant R12. All state points listed in Table 4.1 are noted.

each of the heat exchanger PATs as follows.

Follow the refrigerant from the condenser inlet to the condenser outlet:

$$\mathrm{PAT_{cond}} = \frac{401.0 - 227.6}{\dfrac{401.0-369.8}{337.27} + \dfrac{369.8-242.5}{315.78} + \dfrac{242.5-227.6}{308.6}} = 319\mathrm{K}.$$

Follow the refrigerant from the evaporator inlet to the evaporator outlet:

$$\mathrm{PAT_{evap}} = \frac{228.4 - 356.8}{\dfrac{228.4-351.1}{272.99} + \dfrac{351.1-356.8}{274.5}} = 273\ \mathrm{K}.$$

To obtain the rate of internal entropy production for component k, $S^{\mathrm{k}}_{\mathrm{int}}$, we continue with the approximate scheme of piecewise calculations along each branch. Therefore, each calculation is performed according to the formula:

$$S^{\mathrm{k}}_{\mathrm{int}} = \int_{\mathrm{inlet}}^{\mathrm{outlet}} \mathrm{d}S - \frac{\int_{\mathrm{inlet}}^{\mathrm{outlet}} \mathrm{d}H}{\mathrm{PAT_k}}$$

Applied to each of the four principal chiller components, we have the following results (and remember that specific entropy in kJ kg^{-1} K^{-1} is multiplied by mass flow rate in kg s^{-1} to obtain the *rate* of entropy production in kW K^{-1}):

$$S_{int}^{cond} = \left\{1.095 - 1.638 - \frac{227.6 - 401.0}{319}\right\}(0.084) = 4.82\times10^{-5}\ \text{kW K}^{-1}$$

$$S_{int}^{evap} = \left\{1.580 - 1.10 - \frac{356.8 - 228.4}{273}\right\}(0.084) = 8.123\times10^{-4}\ \text{kW K}^{-1}$$

$$S_{int}^{comp} = (1.6423 - 1.587)\,(0.084) = 4.645\times10^{-3}\ \text{kW K}^{-1}$$

$$S_{int}^{exp} = (1.10 - 1.098)(0.084) = 1.68\times10^{-4}\ \text{kW K}^{-1}$$

$$\text{Rate of total entropy production} = S_{int}^{cond} + S_{int}^{evap} + S_{int}^{comp} + S_{int}^{exp} =$$

$$= 5.67\times10^{-3}\ \text{kW K}^{-1}.$$

Tutorial 4.2

Refer to Tutorial 2.3, both for the properties from standard thermodynamic tables, and for the results of the mass balance and First Law analyses of the 3068 kW single-stage LiBr-water absorption chiller. As noted above, PAT values of sufficient accuracy to estimate internal entropy production needed to be generated with a computer simulation [Chua 1999]. They turn out to be

PAT_{gen} = 370.20 K $\quad$ PAT_{abs} = 323.43 K $\quad$ PAT_{cond} = 319.12 K

PAT_{evap} = 278.15 K.

For the stated thermodynamic conditions at various key locations along the chiller's cycle, determine the rates of internal entropy production at each of the chiller's principal components.

Solution: Recall Figure 2.23 for the state points referred to in the development that follows. The rate of internal entropy production at each component j, S_{int}^{j}, is calculated as follows:

1) **Generator**

$$S_{int}^{gen} = m_6 s_6 + m_1 s_1 - m_5 s_5 - \frac{Q_{gen}}{PAT_{gen}}$$

$$= 1.322\times8.4617 + 16.28\times0.5262 - 17.60\times0.4536 - \frac{4342}{370.2}$$

$$= 0.04075\ \text{kW K}^{-1}.$$

2) **Absorber**

$$S_{\text{int}}^{\text{abs}} = -m_8 s_8 - m_{r2} s_2 + m_3 s_3 + \frac{Q_{\text{abs}}}{\text{PAT}_{\text{abs}}}$$

(where state point 8 here refers to the vapor phase)

$$S_{\text{int}}^{\text{abs}} = -1.322 \times 9.0235 - 16.28 \times 0.31182 + 17.60 \times 0.2438 + \frac{4114}{323.43}$$

$$= 0.00529 \, \text{kWK}^{-1}.$$

3) **Condenser and Evaporator**

Internal dissipation is taken as negligibly small because condensation and boiling are viewed as proceeding in local equilibrium. Hence $S_{\text{int}}^{\text{cond}} = S_{\text{int}}^{\text{evap}} = 0$.

4) **Solution heat exchanger** (SHX)

$$S_{\text{int}}^{\text{SHX}} = m_5 s_5 + m_2 s_2 - m_4 s_4 - m_1 s_1$$
$$= 17.60 \times 0.4536 + 16.28 \times 0.31182 - 17.60 \times 0.2438 - 16.28 \times 0.5262$$
$$= 0.20237 \text{ kW K}^{-1}.$$

5) **Throttling valve**

$$S_{\text{int}}^{\text{valve}} = m_8 s_8 - m_7 s_7 = 1.322 \times 0.67807 - 1.322 \times 0.6385$$
$$= 0.05231 \text{ kW K}^{-1}.$$

$$\text{Total}: \Delta S_{\text{int}} = 0.04075 + 0.00529 + 0 + 0 + 0.20237 + 0.05231$$
$$= 0.3007 \text{ kW K}^{-1}.$$

The solution heat exchanger is responsible for two-thirds of the total internal entropy production. Note that although this irreversibility is primarily due to finite-rate heat transfer, it is internal (rather than external) because it is not involved in the chiller's thermal communication with its environments.

E. DERIVATION OF THE GOVERNING PERFORMANCE EQUATION FOR MECHANICAL CHILLERS

E1. The first two laws of thermodynamics and general modeling of irreversibilities

Consider the steady-state operation of a cyclic mechanical chiller. A schematic is shown in Figure 4.3. In the derivation that follows, refrigerant temperatures at the evaporator and condenser, T_{evap} and T_{cond},

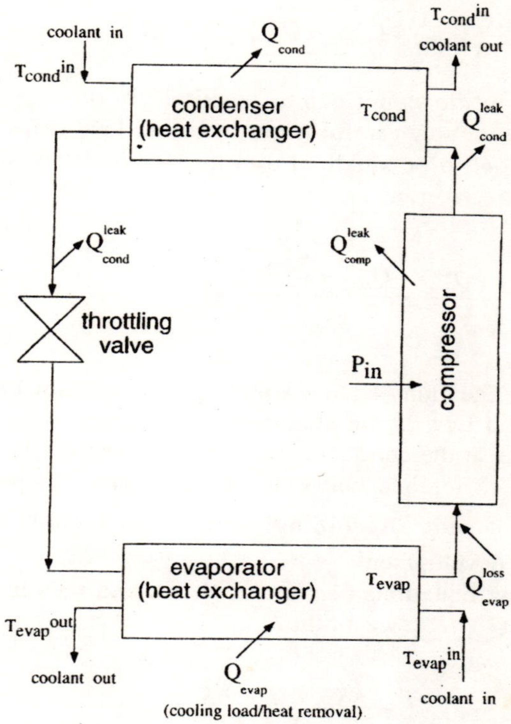

Figure 4.3: Schematic of a mechanical chiller.

respectively, are process-average values. All energy and entropy flows are cycle-average values. Also keep in mind that for thermodynamic state functions such as E and S, the change in a cyclic process is zero. We will take explicit account of: (1) heat transfer rates between the refrigerant and its coolant (Q with the appropriate subscript); (2) heat leaks from the refrigerant to its environment (Q with a superscript "leak" and the appropriate subscript); (3) electrical power input to the compressor (P_{in}); and (4) the rate of internal entropy production ΔS_{int}. The external losses are included *implicitly* by expressing the equations in terms of the refrigerant PATs rather than the coolant temperatures. Explicit account of the external losses, and expressing COP as an explicit function of heat exchanger properties, will be treated in Chapter 5.

From the First Law for the change in the refrigerant's internal energy over one cycle:

$$\Delta E = 0 = Q_{\text{cond}} + Q_{\text{cond}}^{\text{leak}} - Q_{\text{evap}} - Q_{\text{evap}}^{\text{leak}} - P_{\text{in}} + Q_{\text{comp}}^{\text{leak}} \tag{4.15}$$

with all energy flows defined as positive. In our energy and entropy balances, we are now careful about including heat leaks explicitly, even if they turn out to be small. From the Second Law, the entropy balance on the refrigerant is

$$\Delta S = 0 = \frac{Q_{\text{cond}} + Q_{\text{cond}}^{\text{leak}}}{T_{\text{cond}}} - \frac{Q_{\text{evap}} + Q_{\text{evap}}^{\text{leak}}}{T_{\text{evap}}} - \Delta S_{\text{int}} \tag{4.16}$$

(we say that Equation (4.16) invokes the Second Law because one form of the Second Law is the statement that entropy is a state function).

Heat leaks at the condenser are typically negligible due to the relatively small difference between the refrigerant temperature and that of the environment. Accordingly, we do not include the $Q_{\text{cond}}^{\text{leak}}$ term in the ensuing development.

Combining Equations (4.15) and (4.16), and recalling the definition of COP as $Q_{\text{evap}}/P_{\text{in}}$, we obtain

$$\frac{1}{\text{COP}} = -1 + \frac{T_{\text{cond}}}{T_{\text{evap}}} + \frac{T_{\text{cond}}\Delta S_{\text{int}}}{Q_{\text{evap}}} + \frac{T_{\text{cond}}\Delta S_{\text{leak}}}{Q_{\text{evap}}} \tag{4.17}$$

where

$$\Delta S_{\text{leak}} = \frac{Q_{\text{comp}}^{\text{leak}}}{T_{\text{cond}}} + Q_{\text{evap}}^{\text{leak}}\left[\frac{1}{T_{\text{evap}}} - \frac{1}{T_{\text{cond}}}\right]. \tag{4.18}$$

Equation (4.18) indicates that although ΔS_{leak} represents heat leaks from the system to the environment, it can alternatively be interpreted as an internal heat leak from the condenser to the evaporator, and from the compressor to the condenser.

In heat pump mode, with COP defined as $Q_{\text{cond}}/P_{\text{in}}$, it follows simply that

$$\text{COP}_{\text{heat pump}} = 1 + \text{COP}_{\text{chiller}}\left[1 + \frac{Q_{\text{evap}}^{\text{leak}} - Q_{\text{comp}}^{\text{leak}}}{Q_{\text{evap}}}\right]. \tag{4.19}$$

In the limit of negligible heat leaks, Equation (4.19) reduces to the familiar formula in introductory thermodynamics texts

$$\mathrm{COP}_{\text{heat pump}} = 1 + \mathrm{COP}_{\text{chiller}} \,. \tag{4.20}$$

E2. How COP is comprised of contributions from individual classes of irreversibility

Equation (4.17) can be viewed as the sum of individual contributions from 3 types of irreversibilities:

$$\frac{1}{\mathrm{COP}} = \frac{1}{\mathrm{COP}_{\text{endo}}} + \frac{1}{\mathrm{COP}_{\text{int}}} + \frac{1}{\mathrm{COP}_{\text{leak}}} \,. \tag{4.21}$$

The first term on the right-hand side of Equation (4.21) is the contribution of the endoreversible chiller, *i.e.*, accounting solely for losses from finite-rate heat transfer in the heat exchangers. The reservoir temperatures are the coolant inlet temperatures $T_{\text{cond}}^{\text{in}}$ and $T_{\text{evap}}^{\text{in}}$, which are the values the refrigerant temperatures T_{cond} and T_{evap} would reach in the limit of reversible heat exchange. Hence finite-rate heat transfer losses are accounted for implicitly in $\mathrm{COP}_{\text{endo}}$.

For a mechanical chiller, the PATs at the high-temperature side (condenser heat rejection) and low-temperature side (evaporator heat removal) are T_{cond} and T_{evap}, respectively. So another way of writing $\mathrm{COP}_{\text{endo}}$ is

$$\frac{1}{\mathrm{COP}_{\text{endo}}} = -1 + \frac{T_{\text{cond}}}{T_{\text{evap}}} = -1 + \frac{\mathrm{PAT}_{\text{high}}}{\mathrm{PAT}_{\text{low}}} \,. \tag{4.22}$$

In the next chapter, we will translate this expression into one that explicitly accounts for heat exchanger properties and hence can be expressed in terms of coolant temperatures. This extra step is particularly important because coolant temperatures are readily measured non-intrusively, whereas refrigerant temperatures either require problematic intrusive measurements or must be estimated from assumptions about conditions within the refrigerant loop. $\mathrm{COP}_{\text{endo}}$ contains an implicit dependence on Q_{evap}. Once we express COP in terms of coolant temperatures, that dependence will become explicit.

In the contribution of internal losses

$$\frac{1}{\mathrm{COP}_{\text{int}}} = \frac{T_{\text{cond}} \Delta S_{\text{int}}}{Q_{\text{evap}}} \tag{4.23}$$

the PAT for converting internal dissipation into lost input power is T_{cond} (recall that $1/COP = P_{in}/Q_{evap}$).

The heat leak contribution

$$\frac{1}{COP_{leak}} = \frac{T_{cond}\Delta S_{leak}}{Q_{evap}} \tag{4.24}$$

is typically an order of magnitude less than the other terms: small but not negligible for accurate modeling. This claim pertains to commercial mechanical chillers for common air-conditioning and refrigeration uses (and not, for example, for cryogenic applications), and is based on experimental measurements. Representative results to strengthen this point will be presented in Chapter 6. The heat leak term contains a dependence on refrigerant temperatures. But for properly operating commercial chillers, this dependence exerts only a small influence on COP. As we'll see in the case studies of Chapter 6, ignoring this dependence and treating the product $T_{cond}\Delta S_{leak}$ as constant introduces an error that commonly is smaller than the experimental uncertainty in COP.

Tutorial 4.3:

Recall Tutorial 2.2 for a vapor-compression chiller with ammonia as the refrigerant. For both the reversible Carnot cycle and the actual cycle, determine:

(*a*) the process average temperatures at the high and low-temperature ends, PAT_{high} and PAT_{low}

(*b*) the rates of entropy production at the condenser and evaporator, ΔS_{cond} and ΔS_{evap}

(*c*) the rate of internal entropy production ΔS_{int}

(*d*) the COP, calculated with the PATs, and compare with the COP calculated from standard thermodynamic properties in Tutorial 2.2.

The actual cycle has a refrigerant mass flow rate of $m_{refrig} = 0.100$ kg s^{-1}.

Solution: We refer back to all the data provided in Tutorial 2.2 and will not repeat them here. Only the specific additional calculations will be presented.

(*a*) The Carnot cycle

First we calculate the heat transfer and recall the temperatures at the condenser and evaporator (and note that for the reversible cycle, Q_{cond} and Q_{evap} refer to *capacities* rather than *rates*)

$Q_{cond} = h_{2c} - h_{3c} = 1469.9 - 332.8 = 1137.1$ kJ kg^{-1}
$Q_{evap} = h_{1c} - h_{4c} = 1261.7 - 310.6 = 951.1$ kJ kg^{-1}
$T_{cond} = 305.15$ K $\qquad T_{evap} = 255.15$ K .

In the reversible limit

$\mathrm{PAT}_{\mathrm{high}} = T_{\mathrm{cond}}$ and $\mathrm{PAT}_{\mathrm{low}} = T_{\mathrm{evap}}$.

Hence $\Delta S_{\mathrm{cond}} = \dfrac{Q_{\mathrm{cond}}}{T_{\mathrm{cond}}} = 3.73\,\mathrm{kJ\ kg^{-1}\ K^{-1}}$

$$\Delta S_{\mathrm{evap}} = \frac{Q_{\mathrm{evap}}}{T_{\mathrm{evap}}} = 3.73\ \mathrm{kJ\ kg^{-1}\,K^{-1}}$$

and $\Delta S_{\mathrm{int}} = \Delta S_{\mathrm{cond}} - \Delta S_{\mathrm{evap}} = 0$ (as must be the case for a reversible cycle).

COP is calculated from

$$\frac{1}{\mathrm{COP}} = -1 + \frac{\mathrm{PAT}_{\mathrm{high}}}{\mathrm{PAT}_{\mathrm{low}}} + \frac{\mathrm{PAT}_{\mathrm{high}}\,\Delta S_{\mathrm{int}}}{Q_{\mathrm{evap}}} = -1 + \frac{305.15}{255.15} + 0 = \frac{1}{5.103}.$$

The COP here differs (however negligibly) from the previously calculated $\mathrm{COP}_{\mathrm{Carnot}}$ value of 5.11 due to round-off error.

(*b*) The real irreversible cycle

The cooling capacity originally calculated for the real cycle in Tutorial 2.2, Section b, remains unchanged at 1089 kJ kg^{-1}. For the real chiller, we use Q_{evap} and Q_{cond} to denote energy transfer *rates* in kW, and ΔS to denote entropy transfer rates in kW K^{-1}. The cooling rate here is

$$Q_{\mathrm{evap}} = (1089\ \mathrm{kJ\,kg^{-1}})(0.100\,\mathrm{kg\,s^{-1}}) = 108.9\ \mathrm{kW}.$$

The heat rejection rate at the condenser is

$$Q_{\mathrm{cond}} = (h_{3r} - h_{4r})m_{\mathrm{refrig}} = (1745.7 - 332.8)(0.100) = 141.29\ \mathrm{kW}.$$

Applying Equations (4.10) and (4.14) with the thermodynamic properties summarized in Tutorial 2.2, we obtain the PATs:

$$\frac{1}{\mathrm{PAT}_{\mathrm{high}}} = \frac{\dfrac{h_{3r} - h_{4r}}{T_{3r} - T_{4r}}\ln\dfrac{T_{3r}}{T_{4r}} + 2\dfrac{(h_{4r} - h_{5r})}{T_{4r} + T_{5r}} + \dfrac{h_{5r} - h_{6r}}{T_{5r} - T_{6r}}\ln\dfrac{T_{5r}}{T_{6r}}}{h_{3r} - h_{6r}} = \frac{1}{313.1\mathrm{K}}$$

(where the refrigerant specific heat for the sensible-heat pieces is estimated as $\Delta h/\Delta T$, and an arithmetic-average temperature is taken for the phase-change pieces).

$$\text{PAT}_{\text{low}} = \frac{T_{1r} + T_{4r}}{2} = 254.15\ \text{K}$$

$$\Delta S_{\text{cond}} = \frac{Q_{\text{cond}}}{\text{PAT}_{\text{cond}}} = \frac{141.29}{313.1} = 0.4513\ \text{kW K}^{-1}$$

$$\Delta S_{\text{evap}} = \frac{Q_{\text{evap}}}{\text{PAT}_{\text{evap}}} = \frac{108.9}{254.15} = 0.4285\ \text{kW K}^{-1}$$

$$\Delta S_{\text{int}} = \Delta S_{\text{cond}} - \Delta S_{\text{evap}} = 0.02277\ \text{kW K}^{-1}$$

with which we can calculate COP from the chiller characteristic performance equation:

$$\frac{1}{\text{COP}} = -1 + \frac{\text{PAT}_{\text{high}}}{\text{PAT}_{\text{low}}} + \frac{\text{PAT}_{\text{high}}\ \Delta S_{\text{int}}}{Q_{\text{evap}}}$$

$$= -1 + \frac{313.1}{254.15} + \frac{(313.1)(0.02277)}{108.9} = \frac{1}{3.36}$$

which, as expected, agrees with the COP calculated for the same chiller in Tutorial 2.2.

E3. A natural form for chiller characteristic plots

Equation (4.17) implies a convenient form for plotting the chiller's performance curve: 1/COP against 1/(cooling rate). But one must be careful about precisely what is held fixed as cooling rate is varied, and hence under what conditions data for such a plot can realistically be measured. Equation (4.17) is expressed in terms of refrigerant, and not coolant, temperatures. A single curve for a chiller, that would cover the full range of operational cooling rates, would necessitate varying cooling rate at fixed refrigerant temperatures. That would require carefully controlled changes in coolant temperatures while monitoring and controlling refrigerant temperatures - an impractical task.

A far more pragmatic operating strategy is to work at constant coolant temperatures (that can also easily be controlled and measured non-intrusively), and to have the cooling rate vary due to changes in the refrigerant temperatures. In fact, this is how chillers are operated. But it also means that in order to obtain meaningful information from a plot of 1/COP against 1/(cooling rate), we will first have to express refrigerant temperatures in terms of coolant temperatures (via the heat exchanger energy balance equations), and then rewrite Equation (4.17) in terms of coolant temperatures. This exercise will be performed in the following chapter. Examples of characteristic plots, and the type of information that can be gleaned from them, will follow in Chapters 6–10.

F. DERIVATION OF THE PERFORMANCE EQUATION FOR ABSORPTION SYSTEMS

F1. The different modes of absorption machines

Absorption chillers, heat pumps and heat transformers are thermodynamically similar entities. What distinguishes them are the temperature ranges employed, and the useful effect derived from the unit, *i.e.*, cooling, heating or temperature boosting of low-grade heat. The equations governing internal energy and entropy balance are the same; only the definition of the figure of merit (COP) differs. Therefore we proceed with a general derivation. Along the way we will distinguish among different applications.

As with mechanical chillers, our model lumps irreversibilities into 3 categories. For absorption machines, these are, specifically: (1) external losses from finite-rate heat exchange; (2) internal losses from heat and mass transfer in the generator and absorber, pressure drops in the piping, throttling and imperfect regeneration; and (3) heat leaks to or from the environment.

The derivation of the performance equation for absorption devices is similar to that for mechanical chillers, but is complicated by the existence of 4, rather than only 2, reservoirs. There is a generator and absorber, in addition to a condenser and evaporator. Figure 4.4 is a schematic.

F2. Derivation of the characteristic curve for chillers and heat pumps

From the First Law, we have

$$\Delta E = 0 = Q_{\text{cond}} + Q_{\text{cond}}^{\text{leak}} + Q_{\text{abs}} + Q_{\text{abs}}^{\text{leak}} - Q_{\text{evap}} - Q_{\text{evap}}^{\text{leak}} - Q_{\text{gen}} + Q_{\text{gen}}^{\text{leak}} . \quad (4.25)$$

The corresponding Second Law equation is

$$\frac{Q_{\text{cond}} + Q_{\text{cond}}^{\text{leak}}}{T_{\text{cond}}} + \frac{Q_{\text{abs}} + Q_{\text{abs}}^{\text{leak}}}{T_{\text{abs}}} - \frac{Q_{\text{evap}} + Q_{\text{evap}}^{\text{leak}}}{T_{\text{gen}}} - \frac{Q_{\text{gen}} - Q_{\text{gen}}^{\text{leak}}}{T_{\text{gen}}} = \Delta S_{\text{int}} \geq 0 \quad (4.26)$$

where all refrigerant temperatures are process-average values. The

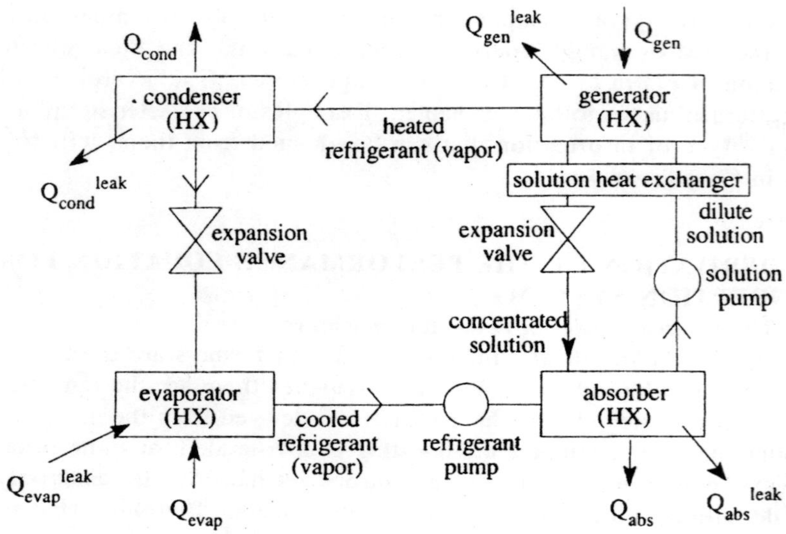

Figure 4.4: Schematic of an absorption chiller.

circulation pumps in absorption units drive saturated liquids through the system. Hence their electrical power consumption is negligible relative to the other energy flows. Experimental data and computer simulation results reveal that heat leaks are small [Chuang & Ishida 1990; Carrier 1962; Abrahamsson 1995].

Heat leaks at the heat rejection reservoirs are negligible because the refrigerant temperature is close to the environmental temperature. Hence we retain only the heat leak terms Q_{evap}^{leak} and Q_{gen}^{leak}. In Chapter 9, we'll evaluate their magnitude and confirm their small influence on system performance.

Let Q_{reject} denote the total heat rejection

$$Q_{reject} = Q_{abs} + Q_{cond} \tag{4.27}$$

and let ξ be the fraction of Q_{reject} that is rejected at the condenser. The fraction ξ introduces an additional control variable (additional compared to mechanical chillers where all heat rejection is effected at the condenser) for optimizing the performance of absorption chillers. Recalling that $COP_{chiller} = Q_{evap}/Q_{gen}$, and combining Equations (4.25)–(4.27), we emerge with the chiller performance formula

$$\frac{1}{\mathrm{COP_{chiller}}} = -1 + \frac{\dfrac{1}{T_{evap}} - \dfrac{1}{T_{gen}} + \dfrac{\Delta S_{int}}{Q_{evap}} + \dfrac{\Delta S_{leak}}{Q_{evap}}}{\dfrac{1}{T_{abs}} - \dfrac{1}{T_{gen}} - \xi\left[\dfrac{1}{T_{abs}} - \dfrac{1}{T_{cond}}\right]} \tag{4.28}$$

where

$$\Delta S_{leak} = Q_{evap}^{leak}\left[\frac{1}{T_{evap}} - \frac{1-\xi}{T_{abs}} - \frac{\xi}{T_{cond}}\right] - Q_{gen}^{leak}\left[\frac{1}{T_{gen}} - \frac{1-\xi}{T_{abs}} - \frac{\xi}{T_{cond}}\right]. \tag{4.29}$$

In analogy to the alternative interpretation of ΔS_{leak} for mechanical chillers in Equation (4.18) at the end of Section E1, ΔS_{leak} for absorption chillers and heat pumps also can be viewed as an internal heat leak. That internal heat leak has two contributions: from the heat rejection reservoirs to the evaporator, and from the generator to the heat rejection reservoirs. In the limit of $1/T_{gen} \to 0$, $\xi = 1$, we have $Q_{gen}^{leak} \to Q_{comp}^{leak}$ and Equation (4.29) reduces to Equation (4.18).

It is straightforward to show that for the heat pump mode, with $\mathrm{COP}_{heat\ pump}$ defined as Q_{reject}/Q_{gen},

$$\mathrm{COP_{heat\ pump}} = 1 + \mathrm{COP_{chiller}}\left[1 + \frac{Q_{evap}^{leak} - Q_{gen}^{leak}}{Q_{evap}}\right] \tag{4.30}$$

so that, in the limit of negligible heat leaks, one recovers the familiar formula Equation (4.20).

F3. Process average temperatures and general expressions for COP

From the arguments developed earlier in the chapter, we can identify the PATs at the high- and low-temperature sides of an absorption chiller. They are entropy-weighted combinations of refrigerant PATs from the contributing reservoirs:

$$\frac{1}{\mathrm{PAT_{high}}} = \frac{1}{T_{abs}} - \frac{1}{T_{gen}} - \xi\left[\frac{1}{T_{abs}} - \frac{1}{T_{cond}}\right] \tag{4.31}$$

$$\frac{1}{\mathrm{PAT}_{\mathrm{low}}} = \frac{1}{T_{\mathrm{evap}}} - \frac{1}{T_{\mathrm{gen}}}. \tag{4.32}$$

The contributions of each of the 3 irreversibility mechanisms to 1/COP are

$$\frac{1}{\mathrm{COP}_{\mathrm{chiller}}} = \frac{1}{\mathrm{COP}_{\mathrm{endo}}} + \frac{1}{\mathrm{COP}_{\mathrm{int}}} + \frac{1}{\mathrm{COP}_{\mathrm{leak}}} \tag{4.33}$$

where

$$\frac{1}{\mathrm{COP}_{\mathrm{endo}}} = -1 + \frac{\dfrac{1}{T_{\mathrm{evap}}} - \dfrac{1}{T_{\mathrm{gen}}}}{\dfrac{1}{T_{\mathrm{abs}}} - \dfrac{1}{T_{\mathrm{gen}}} - \xi\left[\dfrac{1}{T_{\mathrm{abs}}} - \dfrac{1}{T_{\mathrm{cond}}}\right]} = -1 + \frac{\mathrm{PAT}_{\mathrm{high}}}{\mathrm{PAT}_{\mathrm{low}}} \tag{4.34}$$

$$\frac{1}{\mathrm{COP}_{\mathrm{int}}} = \frac{\mathrm{PAT}_{\mathrm{high}}\,\Delta S_{\mathrm{int}}}{Q_{\mathrm{evap}}} \tag{4.35}$$

$$\frac{1}{\mathrm{COP}_{\mathrm{leak}}} = \frac{\mathrm{PAT}_{\mathrm{high}}\,\Delta S_{\mathrm{leak}}}{Q_{\mathrm{evap}}}. \tag{4.36}$$

$\mathrm{PAT}_{\mathrm{high}}$ is the reference temperature for converting internal dissipation into lost work. The reversible Carnot limit (Equation (2.15)) is obtained in the limit of vanishing internal losses, no heat leak losses, and reversible heat exchange, for which

$$\frac{1}{\mathrm{COP}_{\mathrm{chiller}}^{\mathrm{Carnot}}} = -1 + \frac{\dfrac{1}{T_{\mathrm{evap}}^{\mathrm{in}}} - \dfrac{1}{T_{\mathrm{gen}}^{\mathrm{in}}}}{\dfrac{1}{T_{\mathrm{abs}}^{\mathrm{in}}} - \dfrac{1}{T_{\mathrm{gen}}^{\mathrm{in}}}}. \tag{4.37}$$

A practical and implementable chiller performance curve must be expressed in terms of coolant, rather than refrigerant, temperatures. Incorporation of heat exchanger losses for absorption devices will be treated in detail in Chapters 5 and 9.

One can view the mechanical chiller or heat pump thermodynamically as a special case of the absorption chiller in the combined limits: (1) $1/T_{gen} \to 0$ (work, as opposed to heat, input); and (2) $T_{abs} = T_{cond}$ or $\xi = 1$ (only one heat rejection reservoir). The chiller performance characteristic Equation (4.28) for the absorption chiller then reduces to Equation (4.17) for the mechanical chiller; and PAT_{high} and PAT_{low} for the absorption chiller (Equations (4.31)–(4.32)) reduce to those of the mechanical chiller (T_{cond} and T_{evap}, respectively).

F4. Heat transformers

For absorption chillers and heat pumps, we viewed the division of the total heat rejection between the absorber and condenser as a control variable. For the heat transformer, the corresponding control variable is the division of heat *input* between the generator and the evaporator. Let ψ denote the fraction of the total heat input Q_{input} ($Q_{input} = Q_{gen} + Q_{evap}$) that is accepted at the generator. The Second Law equation corresponding to Equation (4.26) is

$$\frac{Q_{cond}}{T_{cond}} + \frac{Q_{abs} + Q_{abs}^{leak}}{T_{abs}} - \frac{(1-\psi)Q_{input} - Q_{evap}^{leak}}{T_{evap}} - \frac{\psi Q_{input} - Q_{gen}^{leak}}{T_{gen}} = \Delta S_{int} \geq 0. \quad (4.38)$$

Again treating the heat leak from the condenser as negligibly small because of the relatively small temperature difference with its environment, we combine Equation (4.38) with Equation (4.25) and the definition that heat transformer $\mathrm{COP} = \dfrac{Q_{abs}}{Q_{gen} + Q_{evap}}$, to obtain

$$\frac{1}{\mathrm{COP}_{\text{heat transformer}}} = \frac{\dfrac{1}{T_{cond}} - \dfrac{1}{T_{abs}} + \dfrac{\Delta S_{int}}{Q_{abs}} + \dfrac{\Delta S'_{leak}}{Q_{abs}}}{\dfrac{1}{T_{cond}} - \dfrac{1}{T_{evap}} + \psi\left[\dfrac{1}{T_{evap}} - \dfrac{1}{T_{gen}}\right]} \quad (4.39)$$

$$\text{where } \Delta S'_{leak} = \frac{Q_{evap}^{leak}}{T_{evap}} + \frac{Q_{abs}^{leak}}{T_{abs}} + \frac{Q_{gen}^{leak}}{T_{gen}} - \frac{Q_{gen}^{leak} + Q_{abs}^{leak} + Q_{evap}^{leak}}{T_{cond}}. \quad (4.40)$$

It is straightforward to identify the individual endoreversible, internal dissipation and heat leak contributions to COP from Equation (4.39). Also, the reversible limit of Equation (2.17) is recovered in the limit of vanishing losses.

G. VALIDITY OF THE CONSTANCY OF INTERNAL LOSSES

A key assumption in accepting the validity of the model proposed here is that the rate of internal dissipation ΔS_{int} is constant over conditions that cover the full range of anticipated chiller operation. Certainly, ΔS_{int} is not rigorously constant for all cooling rates and coolant temperatures. However it appears that ΔS_{int} is constant enough for the full range of actual operating conditions to produce accurate model predictions (accurate meaning comparable to or less than the experimental uncertainty in measuring COP).

Direct and indirect evidence support this assertion. The indirect evidence, conveyed in Chapters 6–10, is the excellent agreement between model predictions and the corresponding experimental results. This assertion pertains to mechanical *and* absorption chillers, and covers an enormous range of cooling rates and operating conditions.

Clearly, however, this is a necessary but not a sufficient piece of evidence. An example of direct and necessary evidence is our having measured the internal entropy production for commercial reciprocating chillers, and having confirmed the validity of the approximation $\Delta S_{int} \approx$ constant [Chua 1995]. It remains for future experimental studies to directly determine the relative constancy of ΔS_{int} for other chiller types and sizes.

At first glance, it may appear unusual that internal dissipation can remain constant as cooling rate is reduced (in some cases the range of cooling rates covers an order of magnitude). A heuristic explanation is that ΔS_{int} is comprised of the product of: (*a*) the entropy production *per unit mass of refrigerant*; and (*b*) the refrigerant mass flow rate. When refrigerant flow rate is reduced, for example by throttling in a reciprocating compressor or by partial closing of the inlet guide vanes in a centrifugal compressor, the entropy production per unit mass *is* increased. It turns out, though, that the decrease in mass flow rate compensates such that the product remains approximately unaltered.

In Chapter 14, we examine cases where ΔS_{int} exhibits a prominent dependence on cooling rate or other key control variables. One well-known case with an exact analytic solution is the thermoelectric chiller. Among mechanical chillers, screw-compressor units exhibit the most noticeable deviation from the constant–ΔS_{int} approximation. In Chapter 14, we'll explain why this is understandable from the special type of internal dissipation in the screw compressor, and we'll illustrate the effect with measured performance data. In addition we'll examine detailed measurements in double-stage absorption chillers where ΔS_{int} deviates noticeably from being constant.

H. PROCESS AVERAGE TEMPERATURE AND EXERGY ANALYSIS

The PAT is the appropriate temperature for translating internal dissipation into the extra work required to generate a given increment in cooling capacity. This is the type of translation or conversion that is critical to chiller manufacturers and to chiller users in selecting chiller components and in evaluating the potential improvements linked to a given change in a compressor, throttler, heat exchanger, absorber, generator, *etc*. The rest of the book contains examples that relate to the interests of chiller manufacturers, chiller designers, cooling engineers and researchers. Chapter 12 is specifically devoted to demonstrating quantitatively how even the internal dissipation in chiller heat exchangers, and hence the accurate calculation of the PAT, can prove significant in diagnostic studies.

Exergy refers to the work that could be produced from a given dissipative process if the dissipated heat could have driven a reversible Carnot heat engine whose cold reservoir is the environment. Setting PAT = T_{env} is only germane if one is performing a broad global optimization typically of interest to the regional energy planner or to the manager of the power plant providing the electricity for a mechanical chiller. Even then, evaluating the work that could be recovered in a reversible power plant is far above the realistic attainable values, even in the most efficient power conversion systems.

Exergy analysis, especially for chillers, is also plagued by the arbitrary and ambiguous definition of T_{env}. Surely a physical variable that can be measured in the laboratory, such as the change in electrical input power to a chiller that follows from a given source of irreversibility, cannot depend upon an arbitrarily-defined reference temperature [Alefeld 1987].

As viewed by the chiller producer and the client purchasing the chiller, exergy losses are irrelevant. However, the work corresponding to the product of internal entropy production and PAT is precisely what interests them since it represents the performance difference for which they are paying directly.

Chapter 5

THE FUNDAMENTAL CHILLER MODEL IN TERMS OF READILY-MEASURABLE VARIABLES

"All exact science is dominated by the idea of approximation." -
Bertrand Russell

A. THE VALUE OF EXPRESSING CHILLER PERFORMANCE IN TERMS OF COOLANT TEMPERATURES

Our aim in this chapter is to express the chiller performance equation in terms of readily-measured coolant temperatures, instead of refrigerant temperatures. Coolant temperatures can be measured non-intrusively and can easily be controlled externally. In contrast, refrigerant temperature measurements usually require intrusive procedures and are problematic to control.

Refrigerant temperatures are easily expressed in terms of coolant temperatures through the heat exchanger energy balance equations. This is exactly what we'll be doing in the following sections. The central pragmatic question is what predictive, diagnostic or optimization capabilities the final result provides.

A problem arises because of the complexity of the final formulae. For example, if the objective is to perform diagnostics, or to be able to predict chiller performance over a wide range of operating conditions from a small number of measurements, then for practical purposes the mathematically cumbersome results are of little help. Nonetheless, in the case studies of Chapters 6–9, we'll show that these unwieldy formulae (in lieu of massive computer simulations) can be used in meaningful chiller optimizations for mechanical and absorption machines. The input to the full inelegant model constitutes independent experimental determination of the model parameters, rather than model parameters arrived at by regression techniques based on a few external measurements. However once the parameters that characterize a given chiller are known, the chiller configuration can be optimized with respect to a number of variables or finite resources of practical interest.

These are problems commonly of substantial interest to chiller manufacturers and designers.

For commercial mechanical chillers, it turns out that approximations can be invoked that convert an unwieldy formula into an equation that is amenable to multiple linear regression. The value and applicability of the fundamental chiller model then extends to the broader user community of cooling engineers and researchers. We will derive an approximate chiller performance equation with which both diagnostic and predictive studies *can* be realized.

For absorption chillers, the combination of the limited format in which manufacturer catalog data are presented, and the inability to reduce the complete chiller performance formula into one which can be handled with simple regression procedures, restricts the value of the fundamental model to optimization studies only. The formulae will be reviewed in this chapter, but the detailed studies are postponed until Chapter 9, where we'll be examining absorption machines (chillers, heat pumps and heat transformers).

B. DERIVATION FOR MECHANICAL CHILLERS

B1. The full expression

To the results derived in the preceding chapter, we add the energy balance relations at the heat exchangers in terms of the coolant inlet or outlet temperatures

$$Q_{\text{cond}} = (mCE)_{\text{cond}}\left(T_{\text{cond}} - T_{\text{cond}}^{\text{in}}\right) = \frac{(mCE)_{\text{cond}}}{1 - E_{\text{cond}}}\left(T_{\text{cond}} - T_{\text{cond}}^{\text{out}}\right) \tag{5.1}$$

$$Q_{\text{evap}} = (mCE)_{\text{evap}}\left(T_{\text{evap}}^{\text{in}} - T_{\text{evap}}\right) = \frac{(mCE)_{\text{evap}}}{1 - E_{\text{evap}}}\left(T_{\text{evap}}^{\text{out}} - T_{\text{evap}}\right) \tag{5.2}$$

where m is coolant mass flow rate; C is coolant specific heat; and E is heat exchanger effectiveness; with m, C and E assumed constant. Heat exchanger effectiveness is the ratio of actual to maximum possible heat transfer rates. It is a dimensionless parameter between zero and unity. The factor E permits us to express the heat transfer equation in terms of the difference between the coolant (inlet or outlet) temperature, and the refrigerant process–average temperature, as in Equations (5.1) and (5.2). The value of E can be calculated for a heat exchanger of known construction, and flows of known m and C values. Formulae for E are tabulated in many texts, *e.g.*, [Kreider & Rabl 1994]. The product mCE

is a heat exchanger's effective thermal conductance (in units of kW K^{-1}).

Tutorial 5.1

For the reader who may not be familiar with the calculation of heat exchanger effectiveness, we offer a simple tutorial. This exercise also highlights the equivalence of a heat exchanger's *mCE* value and its *UA* value (*U* is overall heat transfer coefficient and *A* is overall heat transfer area).

The condenser of a chiller plant comprises a shell-and-tube single-pass counterflow heat exchanger with refrigerant R12 condensing on the shell side. The refrigerant's condensing temperature is 42°C. The coolant (water) flows in the tubes at an average linear speed of 0.90 m s^{-1}, and has a specific heat $C = 4.2$ kJ kg^{-1} K^{-1}. The coolant enters at 30°C and exits at 33°C. The nominal inner and outer diameters of the tubes are 25 and 30 mm, respectively. The length of each tube is $L = 2.85$ m.

Assume that the heat transfer resistances from the refrigerant condensing on the tube walls, and from conductance across the tube walls, are negligible. For evaluating the heat transfer coefficient h_t in the tubes, we use the classic Dittus–Boelter correlation found in standard textbooks on heat transfer for fully developed turbulent flow in tubes [Holman 1992] to obtain $h_t = 3.59$ kW m^{-2} K^{-1}.

(*i*) Determine the overall heat transfer coefficient *U* of the tubes.

(*ii*) Calculate the number of tubes required for the condenser if the total heat transfer rate required is 450 kW.

(*iii*) Calculate the heat exchanger effectiveness.

(*iv*) Calculate the heat exchanger's effective *UA* value.

(*v*) Demonstrate that the heat transfer rate obtained either with *mCE* or *UA* is the same.

Solution:

(*i*) With the given approximations, we only need to account for the contribution of the coolant flow in estimating the overall heat transfer coefficient *U* of a single tube. Hence

$$U = \frac{h_t A_i}{A_0} = 3.59\left[\frac{25}{30}\right]^2 = 2.491 \text{kW m}^{-2}\,\text{K}^{-1}$$

where A_i and A_0 denote the inner and outer surface area of a tube.

(*ii*) The total heat transfer *Q* for *n* tubes is

$$Q = U(n\pi DL)\text{LMTD}.$$

LMTD is the log-mean temperature difference in the heat exchanger, given in terms of the difference ΔT between the refrigerant and coolant temperature at the inlet and outlet (subscripts 1 and 2, respectively). With ΔT_1 = 42–30 = 12 K, and ΔT_2 = 42–33 = 9 K, we have

$$\text{LMTD} = \frac{\Delta T_1 - \Delta T_2}{\ln\dfrac{\Delta T_1}{\Delta T_2}} = \frac{12-9}{\ln\dfrac{12}{9}} = 10.43 \text{ K}.$$

Given that Q is required to be 450 kW, we solve for n as

$$n = \frac{Q}{U(\pi DL)\,\text{LMTD}} = \frac{450}{2.493\,(\pi\, 0.025 \cdot 2.85)\,10.43} = 77.3$$

so we adopt n = 78 tubes in order not to fall short of 450 kW.

(*iii*) The heat exchanger effectiveness E is defined as

$$E = \frac{Q_{\text{actual}}}{Q_{\text{max}}} = \frac{Q_{\text{actual}}}{(mC)_{\text{min}}\,\Delta T_{\text{max}}}.$$

The mC product is far smaller on the coolant side than on the refrigerant side. The coolant's mass flow rate m is

$m = nA_t$ (linear flow speed)(coolant density)

$$= 78\,\frac{\pi(0.025)^2}{4}(0.90)(1000) = 34.5 \text{ kg s}^{-1}.$$

So

$$E = \frac{450}{34.5 \cdot 4.2 \cdot (42-30)} = 0.260.$$

(*iv*) The heat exchanger's effective UA value is estimated from its relation to E [Holman 1992]:

$$UA = -(mC)_{\text{min}} \ln(1-E) = -(34.5 \cdot 4.2)\ln(1-0.260) = 43.4 \text{ kW K}^{-1}.$$

(*v*) To establish the equivalence of the two methods, we calculate

$$Q = (mCE)\Delta T_{\max} = (34.5)(4.2)(0.260)(12) = 452 \text{ kW}$$

and

$$Q = UA(\text{LMTD}) = 43.4(10.43) = 453 \text{ kW}$$

with the two calculations differing by less than 1%, which is commensurate with round-off error.

Now we return to the derivation of the characteristic performance formula for mechanical chillers. We use Equations (5.1) and (5.2) in Equation (4.17), and choose to express COP in terms of coolant *in*let temperatures, to obtain

$$\frac{1}{\text{COP}} = \left\{ \begin{aligned} & -1 + \frac{T_{\text{cond}}^{\text{in}}\,\Delta S_{\text{int}} + Q_{\text{comp}}^{\text{leak}} - Q_{\text{evap}}^{\text{leak}}}{Q_{\text{evap}}} + \left[1 + \frac{Q_{\text{evap}}^{\text{loss}}}{Q_{\text{evap}}}\right] \left[\frac{\dfrac{T_{\text{cond}}^{\text{in}}}{Q_{\text{evap}}}}{\dfrac{T_{\text{evap}}^{\text{in}}}{Q_{\text{evap}}} - \dfrac{1}{(mCE)_{\text{evap}}}} \right] + \\ & + \frac{1 + \dfrac{Q_{\text{evap}}^{\text{leak}} + Q_{s}^{\text{leak}}}{Q_{\text{evap}}} + \dfrac{Q_{s}^{\text{leak}} Q_{\text{evap}}^{\text{leak}}}{Q_{\text{evap}}^{2}}}{(mCE)_{\text{cond}} \left[\dfrac{T_{\text{cond}}^{\text{in}}}{Q_{\text{evap}}} - \dfrac{1}{(mCE)_{\text{evap}}} \right]} + \frac{\Delta S_{\text{int}}}{(mCE)_{\text{cond}}} \left[1 + \frac{Q_{s}^{\text{leak}}}{Q_{\text{evap}}} \right] \end{aligned} \right\} \times$$

$$\times \left[1 - \frac{\Delta S_{\text{int}}}{(mCE)_{\text{cond}}} - \frac{\left[1 + \dfrac{Q_{\text{evap}}^{\text{leak}}}{Q_{\text{evap}}}\right]}{(mCE)_{\text{cond}} \left[\dfrac{T_{\text{evap}}^{\text{in}}}{Q_{\text{evap}}} - \dfrac{1}{(mCE)_{\text{evap}}} \right]} \right]^{-1} \tag{5.3}$$

where for conciseness of notation, we have defined

$$Q_{s}^{\text{leak}} = Q_{\text{evap}}^{\text{leak}} - Q_{\text{cond}}^{\text{leak}} - Q_{\text{comp}}^{\text{leak}}. \tag{5.4}$$

Using typical chiller catalog data to regress for the parameter values in Equation (5.3) is untenable. If, however, each parameter is determined experimentally, then Equation (5.4) can be used to predict how chiller performance changes with each variable, how different allocations of resources such as finite heat exchanger inventory can be optimized, or how an improvement in one component impacts COP. These issues are analyzed in detail in the chapters that follow.

B2. The approximate formula

For commercial mechanical chillers, the actual parameter values permit a marked simplification of Equation (5.3), as shown in [Chua *et al* 1997]. The resulting approximate formula is

$$\frac{T_{\text{evap}}^{\text{in}}}{T_{\text{cond}}^{\text{in}}}\left[1+\frac{1}{\text{COP}}\right]=1+\frac{T_{\text{evap}}^{\text{in}}\Delta \mathbf{S}_{\mathbf{int}}}{Q_{\text{evap}}}+\frac{\mathbf{Q}_{\mathbf{eqv}}^{\mathbf{leak}}\left(T_{\text{cond}}^{\text{in}}-T_{\text{evap}}^{\text{in}}\right)}{T_{\text{cond}}^{\text{in}}\,Q_{\text{evap}}}+\frac{\mathbf{R}Q_{\text{evap}}}{T_{\text{cond}}^{\text{in}}}\left[1+\frac{1}{\text{COP}}\right] \tag{5.5}$$

where the two (condenser and evaporator) heat exchanger thermal conductances combine into a single effective thermal resistance R for the combination of the heat exchangers

$$R=\frac{1}{(mCE)_{\text{cond}}}+\frac{1}{(mCE)_{\text{evap}}} \tag{5.6}$$

and the equivalent heat leak $Q_{\text{eqv}}^{\text{leak}}$ is

$$Q_{\text{eqv}}^{\text{leak}}=Q_{\text{evap}}^{\text{leak}}+\frac{Q_{\text{comp}}^{\text{leak}}\,T_{\text{evap}}^{\text{in}}}{T_{\text{cond}}^{\text{in}}-T_{\text{evap}}^{\text{in}}}. \tag{5.7}$$

Manufacturer catalog data are often reported in terms of $T_{\text{evap}}^{\text{out}}$ instead of $T_{\text{evap}}^{\text{in}}$. In that case, Equation (5.5) remains unaltered, but the parameter R of Equation (5.6) becomes

$$R=\frac{1}{(mCE)_{\text{cond}}}+\frac{1-E_{\text{evap}}}{(mCE)_{\text{evap}}}. \tag{5.8}$$

Equation (5.5) indicates that the chiller can be characterized in terms of 3 parameters that capture the lumped effects of the 3 irreversibilities of internal dissipation, finite-rate heat exchange and heat leaks: ΔS_{int},

R and Q_{eqv}^{leak}, respectively. Furthermore, from common manufacturer catalog data, or from relatively simple non-intrusive in-house measurements, one can apply multiple-linear regression to Equation (5.5) to obtain these 3 parameters. Namely, we emerge with a simple analytic formula that offers predictive and diagnostic capabilities.

Equation (5.5) also reveals the two regimes of chiller operation: (1) external losses dominating at relatively high cooling rate, and (2) internal dissipation prevailing at relatively low cooling rate. The trends noted in Figure 1.4 become transparent:

- at relatively low cooling rates, a plot of 1/COP against 1/(cooling rate), at fixed coolant temperatures, should become a straight line the slope of which yields the internal dissipation;
- at relatively high cooling rates, 1/COP increases rapidly with Q_{evap}; and
- an intermediate point of minimum 1/COP.

B3. Qualifications about the regression fits

A couple of qualifications are in order. First, even if a multiple-linear regression best fit can accurately represent data from a particular chiller, the parameter values may be physically meaningless. Mathematical regression procedures are unrelated to physical realities. For example, ΔS_{int} and R must be positive. The model must not be applied oblivious to physical realities. Therefore it behooves us to demonstrate, with actual chiller data, that the best-fit parameters not only are physically reasonable, but in fact correspond to the variables they purport to represent when those variables are measured independently. That substantiation forms part of the following chapter.

Second, we retain the Q_{eqv}^{leak} in our fits even though it is a small contribution to COP. Neither do we preclude its being negative. As noted in Chapter 4 and as we'll see from chiller data in Chapter 6, the Q_{eqv}^{leak} term is typically an order of magnitude smaller than the other terms: small but non-negligible for accurate modeling. It may also turn out that an installed chiller has developed a significant heat leak that one would like to diagnose in the simple non-intrusive method implied by Equation (5.5). In that case, regressing for Q_{eqv}^{leak} and finding a significant increase relative to its value when the chiller was operating properly is essential. Also, when Q_{eqv}^{leak} is small (the most common case), its value can vary significantly (while remaining small) without noticeably impacting COP. So whether it turns out to be positive is not critical. Furthermore, depending on the component at which the heat leak is occurring, a heat leak can worsen or improve COP. That means that occasionally Q_{eqv}^{leak} is genuinely negative.

C. HEAT EXCHANGER BALANCES FOR ABSORPTION MACHINES

C1. Absorption chillers and heat pumps

The heat exchanger balances noted in Section B are easily applied to the 4 heat exchangers in an absorption machine:

$$Q_j = \pm(mCE)_j\left(T_j - T_j^{\text{in}}\right) = \pm\frac{(mCE)_j}{1-E_j}\left(T_j - T_j^{\text{out}}\right) \tag{5.9}$$

where the subscript j substitutes for generator, condenser, absorber or evaporator; and the ± sign preserves our definition that Q_j be positive. Substituting Equation (5.9) for each of the 4 heat exchangers into Equation (4.28) (*i.e.*, eliminating refrigerant temperatures in favor of coolant temperatures), one obtains a performance equation that is even more ungainly than Equation (5.3), not to mention the fact that more parameters are involved and must be determined in order to implement the model.

We can simplify the analysis slightly by exploiting approximations based on the relative magnitudes of several of the terms, as originally pointed out in [Chua *et al* 1997]. However, a formula as simple as Equation (5.5) does not follow. So even the nominally simplified results here are limited to optimization studies of the type to be presented in Chapter 9. (A quasi-empirical chiller model which offers a limited predictive and diagnostic capability for absorption chillers will be developed in Chapter 10.)

First, for properly-operating commercial absorption units, the heat leaks at the absorber and condenser are negligible. So we retain heat leak terms only at the generator and evaporator. Second, we treat the fraction ξ of total heat rejection Q_{reject} effected at the condenser as a control variable. As we'll see in Chapter 9, one particular partitioning of the heat rejection maximizes system COP.

Third, heat exchange at the generator is usually latent, rather than sensible, so the generator energy balance can be written in terms of the heat exchanger's thermal conductance UA:

$$Q_{\text{gen}} = (UA)_{\text{gen}}\left(T_{\text{gen}}^{\text{in}} - T_{\text{gen}}\right). \tag{5.10}$$

When heat exchange at the generator is dominated by sensible heat, as in water-fired units, Equation (5.9) is retained.

Unlike the heat exchangers in *mechanical* chillers, the heat exchangers in *absorption* chillers can have a non-negligible variation in their UA

value along the heat exchanger. This is caused by the change in solution concentration (and the change in solution temperature that is uniquely linked to the concentration effect) as it traverses the heat exchanger. Rigorous thermodynamic modeling should take this effect into account. However it would probably obviate the possibility of emerging with the types of simple analytic formulae derived here for chiller performance. In the approximate models developed here, we treat the *UA* value of the heat exchangers in absorption machines as constant at their process-average value. For current commercial absorption units, the errors introduced by this approximation are typically of the same magnitude as experimental measurement uncertainties.

Fourth, in many commercial absorption chillers and heat pumps, the absorber and condenser are connected by a single stream of coolant that flows first through the absorber and then through the condenser. In this instance, $T_{\mathrm{cond}}^{\mathrm{in}}$ is additionally constrained by

$$T_{\mathrm{cond}}^{\mathrm{in}} = T_{\mathrm{abs}}^{\mathrm{in}} + \frac{Q_{\mathrm{abs}}}{(mC)_{\mathrm{abs}}} = T_{\mathrm{abs}}^{\mathrm{in}} + \frac{(1-\xi)Q_{\mathrm{reject}}}{(mC)_{abs}}. \tag{5.11}$$

However, the finite capacity $(mC)_{\mathrm{abs}}$ of the coolant stream is occasionally ignored in the analysis. To ensure a meaningful comparison between model and experiment in this case, one simply adopts an infinitely large value for $(mC)_{\mathrm{abs}}$ in the calculation.

C2. Absorption heat transformers

As for absorption chillers and heat pumps, several simplifications can be noted for absorption heat transformers. First, the heat leak from the condenser is usually negligibly small because of the relatively small temperature difference with its environment (and the heat leak at the evaporator assumes a change in sign due to the different mode of operation). Second, a control variable in heat transformer *design* (as opposed to heat transformer *operation* once it is built) is the fraction ψ of the total heat input that is accepted at the generator

$$\psi = \frac{Q_{\mathrm{gen}}}{Q_{\mathrm{gen}} + Q_{\mathrm{evap}}}. \tag{5.12}$$

Third, heat exchange at the generator and evaporator is usually latent, rather than sensible, so that the heat exchanger energy balances can be written analogously to Equation (5.10):

$$Q_{\text{gen}} = (UA)_{\text{gen}}\left(T_{\text{gen}}^{\text{in}} - T_{\text{gen}}\right) \tag{5.13}$$

$$Q_{\text{evap}} = (UA)_{\text{evap}}\left(T_{\text{evap}}^{\text{in}} - T_{\text{evap}}\right) \tag{5.14}$$

$$Q_{\text{abs}} = (UA)_{\text{abs}}\left(T_{\text{abs}} - T_{\text{abs}}^{\text{in}}\right). \tag{5.15}$$

When heat exchange at the absorber is dominated by sensible heat, as in glycol-cooled heat transformers, Equation (5.9) is retained.

C3. Absorption chiller performance curve

By inserting typical realistic values for chiller parameters, one can observe the nature of the characteristic performance curve for absorption machines, as drawn in Figure 5.1. At low values of useful effect, internal dissipation prevails and the curve is linear. This feature is the same as that of mechanical chillers, even though there are additional sources of internal loss. The reason is the approximate constancy of ΔS_{int} over the operating range of interest. At high values of useful effect, external heat exchanger losses dominate and COP decreases rapidly as useful effect is raised. A maximum COP occurs at the point of optimum tradeoff.

Absorption machines designed to exploit waste heat (a nominally free thermal source) tend to be designed so that their rated capacity lies near the point of maximum useful effect. When one pays for the thermal input,

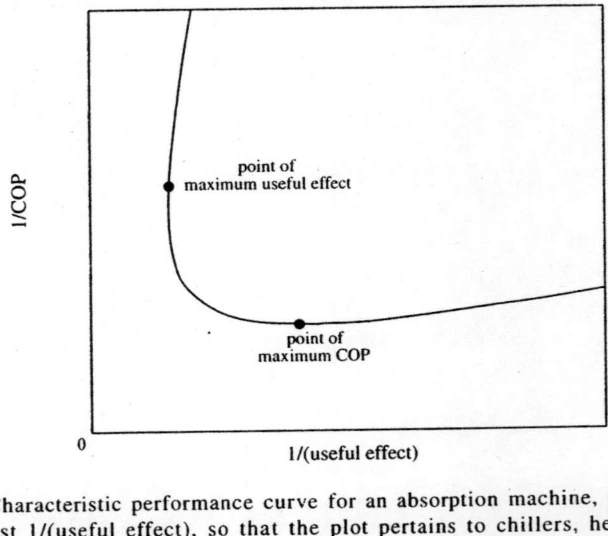

Figure 5.1 Characteristic performance curve for an absorption machine, plotted as 1/COP against 1/(useful effect), so that the plot pertains to chillers, heat pumps and heat transformers. Note the existence of a point of maximum COP and a point of maximum useful effect.

as in gas-fired absorption units, the systems are designed so that the rated capacity falls closer to the point of maximum COP. Perhaps not coincidentally, it appears that manufacturers of gas-fired absorption chillers and heat pumps have empirically evolved designs so that the point of maximum useful effect roughly coincides with that of maximum COP.

A key difference between absorption and mechanical devices is that absorption machines exhibit a measurable point of maximum useful effect. This point exists because absorption systems are driven by a thermal, as opposed to an entropy-less, power source. There are two distinct points (values of COP) for each value of useful effect. The upper branch of the characteristic curve in Figure 5.1 is governed by heat-transfer irreversibilities (*i.e.*, the heat exchange bottleneck) in the generator, so COP decreases as useful effect is lowered. Under realistic conditions, the absorption unit should be designed to operate on the higher-COP branch.

Chapter 6

EXPERIMENTAL VALIDATION OF THE FUNDAMENTAL MODEL AND OPTIMIZATION CASE STUDIES FOR RECIPROCATING CHILLERS

"The fundamental principle of science, the definition almost, is this: the sole test of the validity of any idea is experiment." - Richard P. Feynmann

A. AIMS OF THE CHAPTER

Now that we have derived an analytic model for chiller performance, it is imperative to establish its validity as an accurate predictive tool. Because we contend the model parameters have a clear physical significance, it also behooves us to demonstrate that even if the mathematical form provides good fits to chiller performance data, the best-fit parameters indeed correspond to the physical parameters they are supposed to represent. This consistency check requires data the accuracy and extent of which are ordinarily not found in manufacturer catalogs or journal articles. Two adequate data bases cited below, upon which we will draw for this purpose, are: (1) one publication that reports enough information to test the model; and (2) measurements in our own chiller laboratory.

Once we feel more confident in the model's validity, we will advance to the types of characterization and optimization issues of interest to the designers and producers of chillers. Specifically, in this chapter we will:

(1) test model predictions against experimental data;

(2) examine where the performance of real reciprocating chillers lies along the characteristic performance curves;

(3) optimize chiller design for constrained heat exchanger inventory; and

(4) analyze the performance of highly constrained commercial air-to-air "split" reciprocating chillers in light of this optimization capability. At each step along the way, we'll be examining the wisdom embodied in the empirical evolution of chiller configurations vis-à-vis the thermodynamic optimization calculations we can perform with the analytic model.

B. TEST OF THE FUNDAMENTAL MODEL AS A PREDICTIVE TOOL

B1. Chiller and experimental details

Two independent sets of experimental data are considered. Both data sets pertain to constant coolant flow rates in the chiller's heat exchangers – a situation typical to many installations, and one that will be assumed here and in subsequent chapters unless otherwise specified. The first data set is from a commercial water–cooled reciprocating chiller with a nominal cooling rate of 10.5 kW, tested in our chiller laboratory. Details of the test rig and of the experimental procedures and their accuracy were reviewed in Chapter 3.

Table 6.1 lists 30 sets of experimental measurements that span T^{in}_{evap} = 8–18°C and T^{in}_{cond} = 23.9–35°C, in accordance with Air-conditioning Refrigeration Institute Standard 590-86 [ARI 1986]. Our root-mean-square (rms) error for determining COP experimentally is 2.2%.

The second data set was taken from [Leverenz & Bergan 1983], with a total of 60 experimental points for a water-cooled reciprocating chiller of 70.4 kW nominal cooling rate. These measurements cover a similar range of T^{in}_{evap} and T^{in}_{cond}. The measurements reported are not as extensive as those we performed, so that experimental validation of the thermodynamic model is more limited with this data set than with our laboratory measurements. The rms experimental error for determining COP with this second data set is 5%. The higher rms error probably stems from some non-steady-state measurements and/or poorer instrument accuracy.

B2. Theory versus experiment

Two exercises are performed here. First, we need to ascertain how accurately the simple 3-parameter model of Equation (5.5) can fit actual chiller performance data. Second, in order to demonstrate that the model parameters truly correspond to the physical variables which they purport to represent, we need to compare the best-fit values of these 3 parameters

Table 6.1 Summary of the experimental measurements for the nominal 10.5 kW cooling rate water-cooled reciprocating chiller

Point number	T^{in}_{cond} (°C)	T^{in}_{evap} (°C)	Q_{cond}(kW) ±0.2	Q_{evap}(kW) ±0.2	P_{in} (kW) ±0.01	1/COP ±0.006
1	24.09	8.02	13.8	9.8	3.66	0.373
2	23.89	10.00	14.5	10.5	3.71	0.353
3	23.81	12.37	15.3	11.4	3.81	0.336
4	23.94	13.96	15.6	11.7	3.84	0.329
5	23.90	15.99	16.3	12.4	3.90	0.315
6	23.88	18.00	16.8	12.9	3.92	0.302
7	26.75	8.00	13.7	9.8	3.74	0.380
8	26.75	10.00	14.4	10.5	3.81	0.364
9	26.75	12.41	15.3	11.3	3.94	0.348
10	26.67	14.00	15.9	11.9	3.98	0.335
11	26.74	16.01	16.8	12.7	4.10	0.322
12	26.67	17.99	17.6	13.4	4.14	0.308
13	29.49	8.00	13.3	9.4	3.85	0.408
14	29.38	10.00	14.1	10.1	3.91	0.388
15	29.43	12.39	14.7	10.7	3.98	0.373
16	29.43	13.98	15.6	11.4	4.09	0.357
17	29.39	15.97	16.3	12.1	4.17	0.345
18	29.41	17.99	17.1	12.9	4.22	0.328
19	32.19	7.98	13.2	9.2	3.94	0.428
20	32.24	10.00	14.1	9.9	4.04	0.406
21	32.22	12.40	15.0	10.8	4.16	0.386
22	32.18	13.99	15.6	11.3	4.25	0.375
23	32.22	16.01	16.4	12.1	4.32	0.357
24	32.26	18.02	17.2	12.8	4.43	0.347
25	34.99	8.01	13.0	8.9	4.01	0.450
26	34.88	10.00	13.7	9.5	4.10	0.429
27	35.01	12.40	14.8	10.4	4.26	0.408
28	34.97	13.99	15.4	11.0	4.36	0.396
29	35.05	15.98	16.2	11.7	4.45	0.379
30	34.99	17.99	16.9	12.5	4.55	0.366

from strictly statistical regression fits against their corresponding experimentally-measured values.

For the two experimental data sets noted above, multiple-linear regression is performed to obtain the 3 chiller characteristic parameters:

$\Delta S_{int} = 0.00555$ kW K^{-1}
$R = 2.505$ K kW^{-1} for the 10.5 kW rated chiller
$Q_{eqv}^{leak} = 4.38$ kW

and

$\Delta S_{int} = 0.0366$ kW K^{-1}
$R = 0.127$ K kW^{-1} for the 70.4 kW rated chiller
$Q_{eqv}^{leak} = 26.1$ kW.

How well do these best-fit values account for chiller COP? For the 10.5 kW chiller, the rms error in predicting COP is 0.9%, which is well below the experimental uncertainty of 2.2%. For the 70.4 kW chiller,

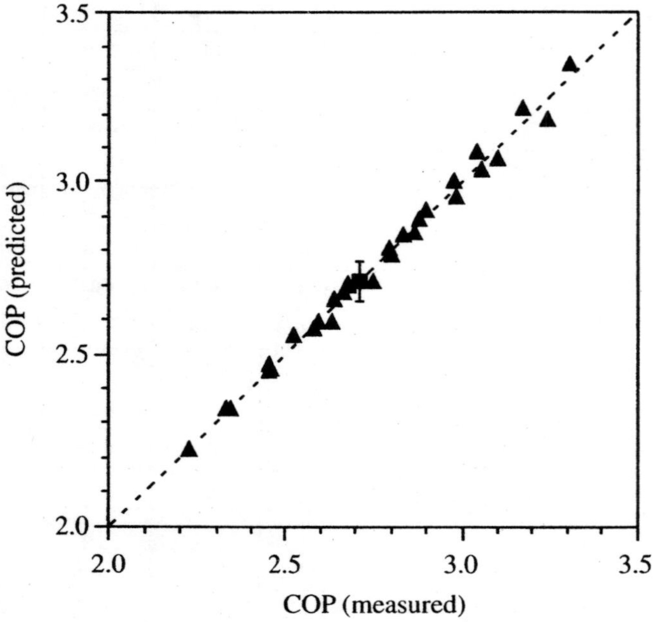

Figure 6.1: Comparison between the model-correlated COP and the corresponding experimentally-measured values for the nominal 10.5 kW chiller tested in our laboratory.

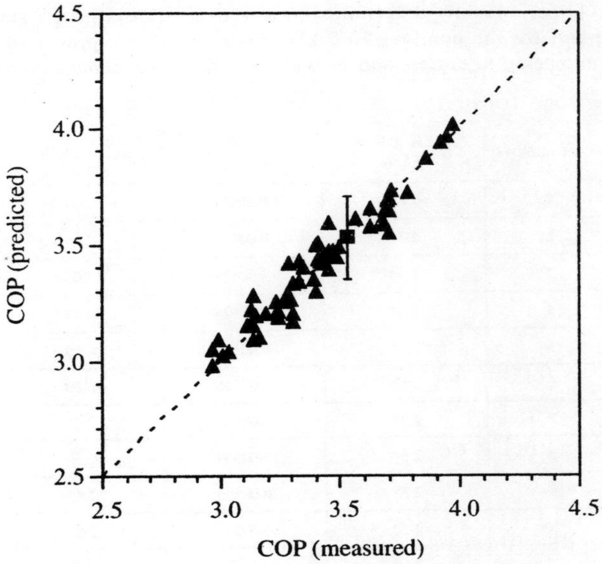

Figure 6.2: Comparison between the model-correlated COP and the corresponding experimentally-measured values for the nominal 70.4 kW chiller described in [Leverenz & Bergan 1983].

the rms error in predicted COP is 1.9%, which is comfortably less than the experimental error of 5%. Figures 6.1 and 6.2 illustrate these points.

Although the data reported in [Leverenz & Bergan 1983] are not sufficiently detailed to permit experimental determination of the 3 parameters at each chiller operating point, the data measured in our laboratory are. The measured parameter values are summarized in Table 6.2. The relevant comparison here is between the regressed values of the 3 model parameters noted above and the experimental values listed in Table 6.2. Also keep in mind that the dominant contributions to COP are from the ΔS_{int} and R terms.

Tutorial 6.1

Relative contributions of the 3 irreversibility sources:

From the best-fit characteristic parameters for the nominal 10.5 kW chiller, determine the relative contributions to chiller performance of internal dissipation, external losses and heat leaks, based on one of the sets of operating conditions listed in Table 6.1.

Table 6.2 Experimentally-determined values of the 3 chiller parameters at each operating point for the nominal 10.5 kW water-cooled reciprocating chiller. The points are numbered to correspond to the same operating conditions listed in Table 6.1.

Point number	R (K kW^{-1}) ±5%	ΔS_{int} (kW K^{-1}) ±0.0003	Q_{eqv}^{leak} (kW) ±0.4
1	2.84	0.00531	2.51
2	2.74	0.00537	2.03
3	2.64	0.00539	3.06
4	2.67	0.00528	4.13
5	2.65	0.00529	4.50
6	2.63	0.00520	6.80
7	2.77	0.00526	2.51
8	2.67	0.00539	2.30
9	2.55	0.00555	2.82
10	2.50	0.00557	2.83
11	2.40	0.00578	2.85
12	2.36	0.00576	2.75
13	2.89	0.00537	2.49
14	2.85	0.00536	2.26
15	2.88	0.00511	3.49
16	2.65	0.00567	1.81
17	2.61	0.00566	2.32
18	2.55	0.00573	1.01
19	2.98	0.00543	2.07
20	2.79	0.00556	2.08
21	2.68	0.00551	2.91
22	2.60	0.00566	3.05
23	2.52	0.00578	2.32
24	2.48	0.00579	2.87
25	3.12	0.00527	2.39
26	2.99	0.00540	2.17
27	2.80	0.00565	2.00
28	2.72	0.00575	2.25
29	2.62	0.00581	2.36
30	2.53	0.00590	2.74

Corresponding model values obtained by multiple-linear regression:
$R = 2.505$ K kW^{-1} $\Delta S_{int} = 0.00555$ kW K^{-1} $Q_{eqv}^{leak} = 4.38$ kW.

Solution: Express Equation (5.5) so that the individual contributions of the 3 irreversibility sources can be compared:

$$\frac{T_{\text{evap}}^{\text{in}}}{T_{\text{cond}}^{\text{in}}}\left[1+\frac{1}{\text{COP}}\right]-1=\frac{T_{\text{evap}}^{\text{in}}\,\Delta\mathbf{S}_{\text{int}}}{Q_{\text{evap}}}+\frac{\mathbf{Q}_{\text{eqv}}^{\text{leak}}\left(T_{\text{cond}}^{\text{in}}-T_{\text{evap}}^{\text{in}}\right)}{T_{\text{cond}}^{\text{in}}\,Q_{\text{evap}}}+\frac{\mathbf{R}Q_{\text{evap}}}{T_{\text{cond}}^{\text{in}}}\left[1+\frac{1}{\text{COP}}\right]. \tag{6.1}$$

The best-fit parameter values are listed above. Select the operating point with

$T_{\text{cond}}^{\text{in}} = 32.19°\text{C}$ $\qquad$ $T_{\text{evap}}^{\text{in}} = 7.98°\text{C}$
$Q_{\text{evap}} = 9.2\ \text{kW}$ $\qquad$ and $\qquad$ $1/\text{COP} = 0.428.$

Then Equation (6.1) yields:

$$\text{total} = \begin{pmatrix}\text{internal}\\ \text{dissipation}\\ \text{term}\end{pmatrix} + \begin{pmatrix}\text{heat}\\ \text{leak}\\ \text{term}\end{pmatrix} + \begin{pmatrix}\text{external}\\ \text{loss}\\ \text{term}\end{pmatrix}$$

$$0.315 = 0.170 + 0.038 + 0.108 = 0.316 \tag{6.2}$$

where in (6.2) the left- and right-hand sides are not identical because we are comparing a best fit against a particular experimental measurement. The relative contributions to the determination of COP are: 54% from internal dissipation, 34% from external losses and 12% from heat leaks.

B3. A qualification: the importance of measurement accuracy

One would imagine that the fundamental model developed here could be used with typical manufacturer catalog data, in the manner presented in Section B2. Specifically, with the usual wide range of reported values of coolant temperatures, cooling rates and COP, we should be able to regress Equation (6.1) to obtain the 3 chiller characteristic parameters. In fact, this is what we have done in the first part of our exercise above.

It appears, however, that the operating points reported in manufacturer catalogs are not all *measured* points. Namely, a few points are measured, and the rest are extrapolated and/or interpolated. This procedure produces nominal data at a level that may be tolerable for sizing air-conditioning systems, but may *not* be acceptable for determining chiller parameters via regression with the fundamental model. Specifically, when we used typical manufacturer catalog data (of the type described in detail in Chapter 10), we found that the best-fit parameters were not physically meaningful. Namely, although an excellent mathematical fit could be obtained, ΔS_{int} could turn out to be negative, or at the very least quite different from physically reasonable values for the rate of internal entropy production. The regressed values of R and Q_{eqv}^{leak} were also far from being physically tenable. This problem can be attributed to the inaccuracy of the nominal data. Hence implementing the fundamental model demands accurate in-house measurements. Tutorial 6.2 highlights this point.

In Chapter 10, we'll develop a quasi-empirical analytic chiller model that is more robust and can work with the nominal data from manufacturer catalogs. The price paid, however, will be the inability to identify the parameters that characterize the chiller with specific irreversibility mechanisms, and hence the loss of an optimization capability.

Tutorial 6.2

Recall the chiller reported in Section B1 and the associated 30 data points tabulated in Table 6.1. In this tutorial, we'll demonstrate the price paid for producing nominal data by extrapolation from a few measured points.

We select 4 characteristic operating conditions from Table 6.1: points 8, 9, 14 and 15.. Then we create the remaining 26 nominal data points in the following manner. Via multiple linear regression (in a standard PC spreadsheet program), we empirically fit the input power P_{in} as a linear function of T_{cond}^{in}, T_{evap}^{in} and Q_{evap}. Then, based on this best-fit linear relation, we calculate P_{in} at the coolant temperatures and cooling rates of the remaining 26 points. The resulting set of 30 data points (comprised of 4 actual measured points and 26 extrapolated points) is then used in the multiple-linear regression calculation prescribed and illustrated in Section B2, to generate the 3 characteristic chiller parameters: ΔS_{int}, R and Q_{eqv}^{leak}.

The results of this exercise are

based on the extrapolated nominal data set	based on the actual 30-point data set (for comparison)
$\Delta S_{int} = -0.00390$ kW K^{-1}	$\Delta S_{int} = 0.00555$ kW K^{-1}
$R = 5.556$ K kW^{-1}	$R = 2.505$ K kW^{-1}
$Q_{eqv}^{leak} = 23.66$ kW	$Q_{eqv}^{leak} = 4.38$ kW

with the rms error for correlating COP being well below the experimental error in both cases. Clearly, however, not only are the best-fit parameters inaccurate, but the negative value of ΔS_{int} is inadmissible.

C. WHERE ACTUAL CHILLER PERFORMANCE LIES ON THE CHARACTERISTIC CURVE

The analytic chiller model provides a window through which we can view where chillers actually operate along their characteristic performance curves. Put another way, how close to the point of maximum COP is a chiller's rated operating condition? Have chiller manufacturers empirically evolved configurations that are as efficient as possible for prescribed cooling needs?

The comprehensive and accurate data set summarized in Tables 6.1 and 6.2 for a small typical commercial reciprocating chiller is well suited to the task. In Figure 6.3, we plot the characteristic performance curve for this nominal 10.5 kW chiller at 5 different operating points that cover the nominal rated condition plus conditions for which the curves lie well above and below the rated condition curve. These correspond to points 3, 6, 15, 19 and 25 in Tables 6.1 and 6.2.

We see the basic performance features in Figure 6.3: (*a*) a linear regime at the lower cooling rates, where chiller behavior is dominated by internal losses; (*b*) a region at higher cooling rates where COP changes rapidly with cooling rate, and finite-rate heat transfer is the key bottleneck; and (*c*) a point of maximum COP at the optimal balance between internal dissipation and heat transfer losses. For many chiller

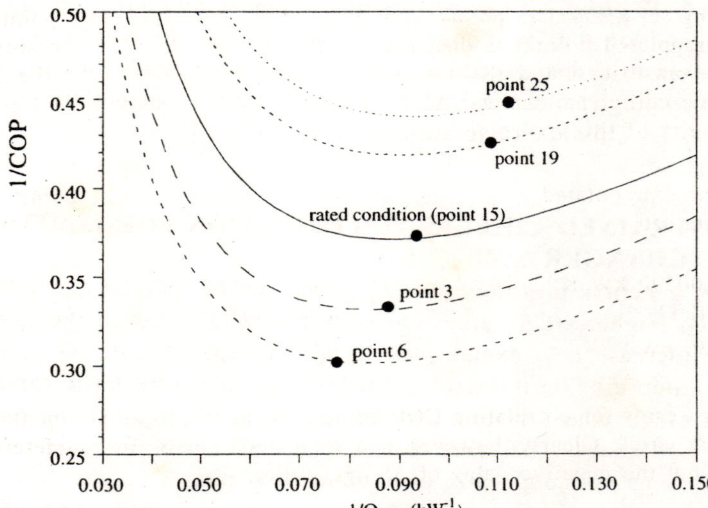

Figure 6.3: Plots of 1/COP against 1/(cooling rate) for the nominal 10.5 kW water-cooled reciprocating chiller for which performance data are summarized in Tables 6.1 and 6.2. Each of the 5 measured points is indicated by a ●.

types, such as centrifugal, screw-compressor, thermoelectric and others, much of the characteristic curve can be accessed experimentally. For reciprocating chillers, however, once one fixes the coolant temperatures (*i.e.*, the reservoir temperatures), only a single point can be measured for a given (theoretical) curve. For different coolant temperatures, each performance point belongs to a different curve. The curves correspond to varying cooling rate by changing refrigerant temperatures (at fixed coolant temperatures). They are a hypothetical construct that illustrates the nature of chiller operation and shows where an actual operating point lies relative to maximum-COP operation. The chiller performance curves in Figure 6.3 are calculated with Equation (6.1) and experimentally-determined values of the 3 chiller characteristic parameters. The fact that each data point lies exactly on its predicted curve is not a test of a theoretical prediction. Rather it is simply a confirmation of the accuracy of the experimental measurements and the fact that all energy flows have been accounted for in Equation (4.15). It is the entropy-balance Equation (4.16) (and the final result of Equation (5.3), of which Equation (6.1) here is an approximate form) that determines where along the chiller performance curve an operating point lies. The curves in Figure 6.3 are characteristic of the empirical wisdom embedded in the evolution of commercial reciprocating chillers. Specifically, they were developed so that their nominal maximum-cooling-rate operating point is around the maximum COP point. The extremum is a broad one. Around the maximum-COP point, chiller performance is more tolerant to changes in cooling rate on the low-cooling-rate side. Hence one would expect chiller design to accommodate a range of operating conditions below maximum cooling rate that fall to the right of the maximum COP point in Figure 6.3. This is exactly what is observed.

D. CONSTRAINED CHILLER OPTIMIZATION FOR LIMITED HEAT EXCHANGER SIZE

The chiller's thermal inventory ($(mCE)_{cond}$ and $(mCE)_{evap}$) is an expensive commodity. Higher mCE values ensure higher COPs, but at the same time they increase heat exchanger size and pumping costs. As noted above, commercial reciprocating chillers appear to be built for an operating range where internal losses are balanced against the heat exchange bottleneck. Therefore chiller COP will not be insensitive to changes in heat exchanger size or coolant flow rates.

To illustrate how the thermodynamic model can be used to determine the component parameters for maximum-COP performance when practical cost constraints are introduced, we consider the chiller's total thermal inventory as a design constraint, namely

$$(mCE)_{cond} + (mCE)_{evap} = (mCE)_{total} = \text{constant}. \tag{6.3}$$

Fixed heat exchanger inventory can be represented by a number of variables that are not rigorously equivalent, *e.g.*, total thermal conductance, total heat exchange area, and total *mCE*, among others. The qualitative trends in the predictions we're about to make are the same for each of these choices, although the quantitative results can vary. Which variable will be selected can be manufacturer and device-specific. There have been earlier studies of the impact of the constraint of fixed total heat exchanger thermal conductance or fixed total heat exchanger size, but no comparisons with actual chiller constructions were attempted.

In determining the chiller's maximum COP operating conditions, there are two degrees of freedom, which we select as Q_{evap} and $(mCE)_{evap}$. Two equations must then be solved simultaneously:

$$\frac{\partial\,\text{COP}}{\partial\,Q_{evap}} = 0 \quad \text{and} \tag{6.4}$$

$$\frac{\partial\,\text{COP}}{\partial\,(mCE)_{evap}} = 0 \tag{6.5}$$

(applied to Equation (5.3)).

As a representative example to examine chiller optimization for constrained total heat exchanger inventory, we take chiller data from the experimental study of [Liang & Kuehn 1991]. The experimental measurements were accurate and extensive enough to allow determination of all the chiller parameters in Equation (5.3). The chiller is a nominal 7.56 kW cooling rate water-cooled commercial reciprocating unit. Its basic characteristics are listed in Table 6.3. A description of the calculation of the chiller parameters from experimental measurements and thermodynamic tables was detailed in [Chua *et al* 1996].

This chiller's characteristic performance curve at the rated operating condition is shown in Figure 6.4. The thermal lift at the rated operating condition is almost zero, which means that the chiller's cooling rate is essentially the highest possible for that particular machine. Most of the time, the chiller would operate at lower cooling rates, and hence at points on Figure 6.4 that lie to the right of the indicated experimental point.

Let's assume that the total heat exchanger inventory currently installed is the fixed value: $(mCE)_{total}$ = 0.926 kW K^{-1}. The questions are: (a)

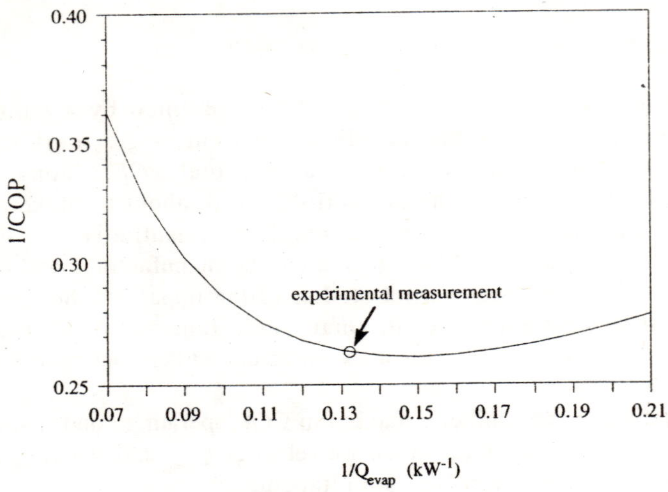

Figure 6.4: Characteristic chiller performance curve for the nominal 7.56 kW water-cooled reciprocating chiller, including the measurement at its rated condition.

how best to divide $(mCE)_{total}$ between the evaporator and the condenser; (b) for that nominally optimal allocation, what is the cooling rate that maximizes COP; and (c) what is the globally maximum COP. From the experimentally-determined chiller parameters, and applying Equations (6.3)–(6.5) to the governing Equation (5.3), we can generate these results. They too are listed in Table 6.3, and appear to confirm that commercial chiller design has evolved toward the conditions that emerge from a constrained thermodynamic optimization.

E. HIGHLY CONSTRAINED OPTIMAL DESIGNS: AIR-COOLED SPLIT RECIPROCATING CHILLERS

For water-cooled chillers, both the evaporator and condenser heat exchangers are situated in the same location, with little if any space constraints (beyond the cost of the heat exchanger). A practical and more severely constrained device is the commercial air-to-air "split" chiller unit (see Figure 6.5). The condenser and compressor are placed outdoors. The evaporator (direct expansion cooling coil) is housed in the indoor unit, where space is often at a premium. When the indoor occupancy space is a major constraint, the chiller designer would impose the evaporator's thermal inventory (rather than the total thermal inventory) as a constraint. Namely, the size of the fan coil would be selected, and the design cooling capacity would then be modified to a nominally sub-

Table 6.3 Experimentally-measured chiller properties and calculated optimal operating conditions for the nominal 7.56 kW water-cooled reciprocating chiller, at standard chiller rating conditions.

T^{in}_{cond} (°C)	9.33
T^{in}_{evap}(°C)	9.28
P_{in} (kW)	1.99
Q^{leak}_{evap}(kW)	0.14
Q^{leak}_{cond} (kW)	0.15
$Q^{optimal}_{evap}$ (kW)	0.08
ΔS_{int} (kW K^{-1})	0.00279
$(mCE)_{cond}$ (kW K^{-1})	0.463
$(mCE)_{evap}$ (kW K^{-1})	0.463
Q_{evap} (kW)	7.56
1/COP	0.263
calculated global optimum (from the thermodynamic model)	
$(mCE)^{optimal}_{evap}$(kW K^{-1})	0.439
$Q^{optimal}_{evap}$(kW)	6.71
$(1/COP)^{optimal}$	0.261

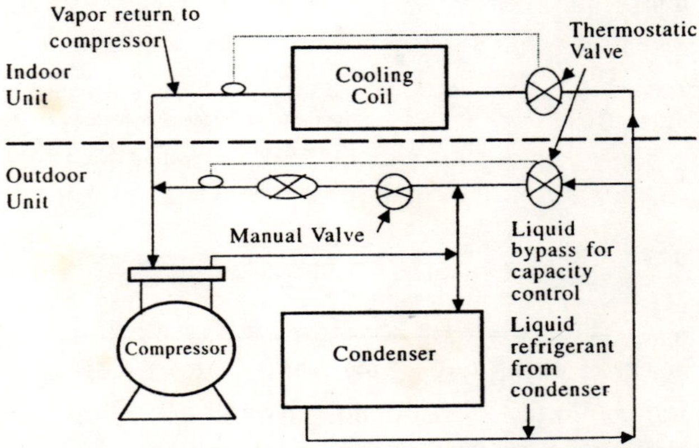

Figure 6.5: Schematic of a split air-air reciprocating chiller. The broken line indicates the remote location of the evaporator cooling coil. Both the evaporator and condenser are air cooled. For capacity control, the hot refrigerant gas or liquid refrigerant from the condenser can be routed to bypass the evaporator, and returned directly to the compressor.

optimal value in order to accommodate the new size limitation. The issue then is the determination of optimal operating conditions for this differently-constrained situation.

In a special publication on air-to-air split cooling units [Toyo 1989], sufficient experimental data are reported for 11 reciprocating chillers to permit evaluation of the key chiller variables, some of which are listed in Table 6.4. The exercise reduces to calculating the cooling rate at which COP is maximized, and the value of that maximum COP. We use Equation (5.3) with the chiller parameters determined from the data presented in [Toyo 1989] to generate the chiller's characteristic performance curve, and then identify the point of maximum COP. These results, listed in Table 6.4, relate to rated conditions of T_{evap}^{in} = 35.0°C and T_{evap}^{in} = 27.0°C (the rated coolant temperatures are higher than in earlier examples due to the lower thermal inventory for split chillers and fixed cooling rate requirements). In Table 6.4, note the striking proximity of actual device performance to nominally optimal behavior.

To sharpen the perspective on chillers with a highly constrained heat exchanger inventory, we plot in Figure 6.6 six characteristic performance curves for one of these split chillers. The location of the actual operating point along each curve again points to the view that commercial reciprocating chillers are tailored to approximately maximum-COP

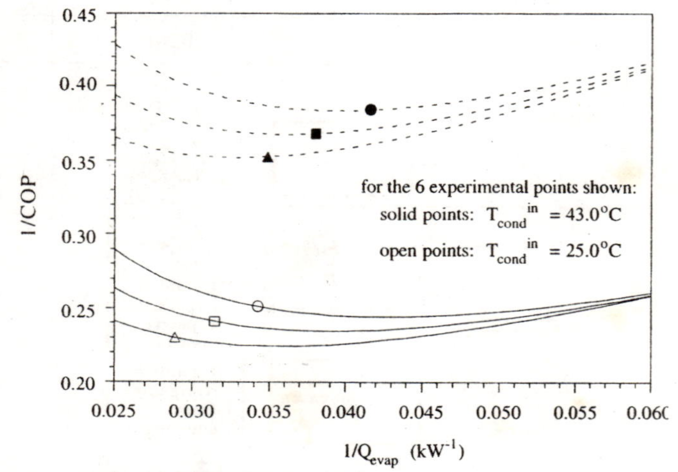

Figure 6.6: Characteristic chiller plots for 6 sets of coolant temperatures for one of the split-unit air-cooled reciprocating chillers noted in Table 6.4. Each measured point lies on a separate curve which is calculated based on experimental measurements and the analytic chiller model. Note that chiller operation falls more to the heat-exchanger-dominated side (*i.e.*, to the left) of the maximum-COP point than for the water-cooled chillers analyzed earlier.

Table 6.4: Comparison between model predictions of optimal operating conditions and experimental performance data, for commercial air-to-air split chillers, at rated coolant temperatures $T^{in}_{cond} = 35.0°C$ and $T^{in}_{evap} = 27.0°C$. In particular, compare the values of measured and predicted Q_{evap} and COP for each chiller.

Model No.	38PE008	38PE009	38PE010	38PE012	38PE015	38PE018	38PE020	38PE025	38PE030	38PE036	38PE045
computed and measured data from [Toyo 1989]:											
ΔS_{int} (kW K^{-1}) computed	0.012	0.013	0.016	0.013	0.015	0.022	0.029	0.025	0.028	0.043	0.066
$(mCE)_{evap}$ (kW K^{-1}) computed	0.781	1.021	1.101	1.224	1.672	2.048	2.180	2.422	4.772	4.055	4.898
$(mCE)_{cond}$ (kW K^{-1}) computed	1.739	2.053	2.389	2.649	2.925	3.731	4.623	4.975	5.167	7.536	9.956
Q_{evap} (kW) measured	18.6	23.3	26.0	29.1	36.6	46.5	52.3	58.1	73.3	93.0	116.0
1/COP measured	0.368	0.345	0.361	0.306	0.303	0.323	0.348	0.309	0.271	0.319	0.356
constraint: fixed *evaporator* thermal inventory. optimal operating conditions for maximum COP											
Q_{evap} (kW)	20.6	24.6	28.3	27.4	34.2	45.1	54.1	53.4	71.6	88.8	121.9
1/COP	0.366	0.344	0.360	0.306	0.302	0.323	0.348	0.308	0.271	0.318	0.356

operating conditions. However in air-cooled chillers, one expects heat exchangers to pose a larger bottleneck effect than in water-cooled chillers. This effect would manifest itself in chillers operating further to the left on the characteristic chiller plot, while still straddling the maximum-COP conditions. This too appears to be confirmed by the data-based results.

Chapter 7

FINITE-TIME THERMODYNAMIC OPTIMIZATION OF REAL CHILLERS

"To start in a hurry
and finish in haste
will minimize worry
and maximize waste."
Piet Hein

A. GLOBAL OPTIMIZATION WITH RESPECT TO FINITE TIME AND FINITE THERMAL INVENTORY

This chapter is devoted to investigating the best way to allocate the finite cycle time of a mechanical chiller among its principal components. Namely, for a given fixed cycle time, what should the residence time of the refrigerant be in the compressor, expansion device, evaporator and condenser in order to maximize COP for a prescribed cooling rate. From the perspective of chiller manufacturers, the relative residence time in each chiller component can be viewed as a control variable.

This assertion entails an unorthodox view of precisely what constitutes a control variable at the stage of chiller design. We contend that whether a manufacturer recognizes or treats relative residence time as a control variable is not at issue. The very fact that relative residence time introduces an additional degree of freedom in the design (as opposed to the operation) of the chiller is a sufficient incentive to explore how chiller performance can be ameliorated with respect to it.

We emphasize that the type of optimization considered in this chapter ceases to exist the moment the chiller components have been selected. From the viewpoint of chiller installers and consumers, no optimization is implied. Neither does the optimization exercise exist if the chiller manufacturer must accept specific off-the-shelf components.

In material terms, the control variable contemplated here is actually the relative refrigerant charge of each component. At issue is exactly how the total refrigerant mass is distributed among the condenser, evaporator, compressor and expansion device. For a chiller operating at constant refrigerant mass flow rate, at steady state, the relative refrigerant charge in any component is identical to the relative residence time the refrigerant spends in that component (mass = {constant mass flow rate} · time). Relative residence time refers to the actual time the refrigerant resides in a given component, relative to the total cycle time.

The chiller may possess an accumulator that contains a substantial quantity of refrigerant, but accumulators serve to accommodate transient operation. At steady state, essentially all the refrigerant mass is accounted for by the 4 principal components noted above.

Chiller manufacturers may commonly characterize heat exchangers, for example, in terms of their overall *UA* values and the mass flow rates traversing them. The fluid volume may not be considered as a design variable. In this exercise, however, we take a step back in the design process, and treat the heat exchanger's *mCE* product per unit mass of refrigerant as a valid control variable. This is the extra degree of freedom introduced here. Clearly, when the heat exchanger is viewed as one of several components in a chiller cycle, this control variable can be expressed as the component's relative refrigerant charge, for a given total refrigerant charge in the chiller.

Rather than continuing to refer to the new control variable as relative refrigerant charge, and because it is rigorously equivalent to relative residence time (for a fixed chiller cycle time or, equivalently, a fixed mass flow rate and fixed total charge), we shall call this additional degree of freedom relative residence time, and denote it by the symbol Ξ_i for component i. Also, in light of the evolution of the discipline of finite-time thermodynamics during the past 25 years, we offer this analysis as a relatively simple but bona fide example of optimizing the thermodynamic performance of real machines with respect to how a given finite time should be apportioned among the device's elements.

In Chapter 6, we explored optimizing a chiller when its heat exchanger inventory is constrained. Here we broaden the optimization to include the finite resource of time, and modify exactly what fixed heat exchanger inventory signifies. The global optimum with the additional control variable of time will now be determined. One benefit of using the thermodynamic model developed in the previous chapters is obtaining analytic results for optimal time divisions and optimal specific (per unit refrigerant charge) heat exchanger allocation. In developing an analytic predictive chiller model in earlier chapters, we treated a chiller as a sort of input–output device, viewed from the outside and probed only with externally-measurable parameters such as input power, cooling rate and coolant temperatures. Here, we must intrude into the compressor, throttler, condenser and evaporator, because we need to quantitatively characterize the dissipation in each component in order to perform the finite-time optimization. This is one reason why the type of extensive chiller measurements needed for such a study is not common.

Using actual chiller performance data, we'll see that the design and construction of commercial reciprocating chillers have evolved to the optimal operating strategies calculated from finite-time thermodynamics. This reflects the empirical wisdom embodied in these constructions. (Since both finite time and finite heat exchanger inventory are the constraints of practical interest here, a more appropriate rubric might be finite-*resource* thermodynamics; but we retain the finite-time appellation for historical reasons.)

We will also show that, for the particular set of constraints that relates to practical designs for manufacturers, maximizing COP is equivalent to minimizing entropy production in the universe (and not just inside the chiller). This point is not trivial because maximum COP and minimum dissipation in the universe (*i.e.*, the combination of the chiller and its surroundings) are not necessarily identical objectives.

B. HOW FINITE TIME ENTERS GOVERNING PERFORMANCE EQUATIONS

Let's revisit the derivation presented in Chapter 4 for the chiller's thermodynamic performance.

The energy balance, Equation (4.15), remains unaltered. Consideration of the relative residence time of the refrigerant in each of the principal chiller components enters in the entropy balance. Equation (4.16) is modified as follows:

$$\frac{Q_{\text{cond}} + Q_{\text{cond}}^{\text{leak}}}{T_{\text{cond}}} - \frac{Q_{\text{evap}} + Q_{\text{evap}}^{\text{leak}}}{T_{\text{evap}}} - \delta S_{\text{comp}} \Xi_{\text{comp}} - \delta S_{\text{exp}} \Xi_{\text{exp}} = 0 \tag{7.1}$$

where δS_{comp} = entropy production rate per relative residence time during compression; δS_{exp} = entropy production rate per relative residence time during expansion/throttling; Ξ_{comp} = fraction of cycle time for compression (dimensionless); and Ξ_{exp} = fraction of cycle time for expansion.

The relative residence times Ξ will be viewed as control variables in the design of optimized chillers. Equivalently, we ask what the optimal allocation of cycle time among the various branches of the refrigerant cycle is. Again, we hasten to stress that varying the relative time allocation does not relate to the operation of a chiller that has already been built. Rather, we are adopting a pre-construction perspective of the chiller. After the chiller has been assembled, the key degrees of freedom considered here cease to exist.

The rate of internal entropy production ΔS_{int} is given by:

$$\Delta S_{\text{int}} = \delta S_{\text{comp}}\, \Xi_{\text{comp}} + \delta S_{\text{exp}}\, \Xi_{\text{exp}}. \tag{7.2}$$

We relate to the compression and throttling branches as adiabatic, which is an excellent approximation for throttling. Non-adiabaticity for compression is generally small, with the degree of deviation being reflected in the low experimental value of $Q^{\text{leak}}_{\text{comp}}$.

We also assign all the internal dissipation to the compressor and expansion device. In accordance with common practice in chiller analysis, internal losses in the heat exchangers are treated as negligible or lumped with the internal losses in the other components. In Chapter 12, we'll show that internal dissipation in heat exchangers is not always negligible, and can noticeably impact diagnostic procedures. For the procedures outlined here, however, the internal dissipation in the heat exchangers of the mechanical chillers has only a small impact on the optimization and therefore is omitted.

Since it is coolant, rather than refrigerant, temperatures that are readily and non-intrusively measurable (as well as heat flows and power input), and in terms of which chiller performance equations should conveniently be cast, the energy balance on the heat exchangers is also expressed in terms of relative residence times:

$$Q_{\text{cond}} = \Xi_{\text{cond}}(mCE)'_{\text{cond}}\left(T_{\text{cond}} - T^{\text{in}}_{\text{cond}}\right) \tag{7.3}$$

$$Q_{\text{evap}} = \Xi_{\text{evap}}(mCE)'_{\text{evap}}\left(T^{\text{in}}_{\text{evap}} - T_{\text{evap}}\right) \quad . \tag{7.4}$$

where Ξ_{cond} = fraction of cycle time in the condenser; and Ξ_{evap} = fraction of cycle time in the evaporator. Note that the products mCE with a prime (′) superscript are essentially values of mCE per relative refrigerant charge, and are equivalently viewed here as mCE values per relative residence time. The heat exchangers can be viewed as autonomous components characterized by $(mCE)'$ values and by design control variables Ξ. The same observation pertains to the compressor and throttler for the local entropy production rates δS and the relative residence times Ξ. The tradeoff between Ξ_{cond} and Ξ_{evap} is a principal optimization step here in chiller design. By definition, the relative residence times are normalized – the finite-time constraint:

$$\Xi_{comp} + \Xi_{exp} + \Xi_{cond} + \Xi_{evap} = 1. \tag{7.5}$$

We do not need to know the actual values of the residence times or even the cycle time itself to complete the optimization exercise. The relative residence times Ξ are sufficient. We now combine Equation (4.15) with Equations (7.1)–(7.5) above to yield the characteristic chiller curve for 1/COP as a function of $1/Q_{evap}$ and all key chiller variables:

$$\frac{1}{\mathrm{COP}} + 1 = \frac{\dfrac{T_{cond}^{in}\Xi_{cond}(mCE)'_{cond}}{Q_{evap}}\left[\Delta S_{int} - \Xi_{evap}(mCE)'_{evap} - \dfrac{T_{evap}^{in}\Delta S_{int}\Xi_{evap}(mCE)'_{evap}}{Q_{evap}}\right]}{\Xi_{cond}(mCE)'_{cond} + \Xi_{evap}(mCE)'_{evap} - \Delta S_{int} + \dfrac{T_{evap}^{in}\Xi_{evap}(mCE)'_{evap}}{Q_{evap}}\left[\Delta S_{int} - \Xi_{cond}(mCE)'_{cond}\right]}. \tag{7.6}$$

In deriving Equation (7.6), we have neglected heat leaks. The usually small heat leaks exert a negligible influence on the optimal allocation of residence time and heat exchanger inventory. The derivation in the absence of heat leaks also results in simple analytic formulae with which fundamental functional dependences are transparent and easily evaluated.

C. PERFORMING THE GLOBAL OPTIMIZATION

The global optimization we consider is maximizing COP as a function of 3 control variables, subject to two constraints. The control variables are the relative residence times, the partition of specific heat exchanger inventory, and cooling rate. The constraints are fixed cycle time (Equation (7.5)) and fixed heat exchanger inventory per refrigerant charge (*i.e.*, per relative residence time)

$$(mCE)'_{cond} + (mCE)'_{evap} = (mCE)'_{total} = \text{constant} \equiv C_0. \tag{7.7}$$

For a given compressor and expansion device and for realistic operating conditions, COP is a monotonically decreasing function of Ξ_{comp} and Ξ_{exp} so that their optimal values are in principle zero.

Fundamental limits to the operation of compressors and expansion devices restrict Ξ_{comp} and Ξ_{exp} to certain minimum values commensurate with typical required chiller cycle times. Therefore the optimal Ξ_{comp} and Ξ_{exp} values are the minimum values consistent with current technology. Examination of the operating characteristics of the throttlers and reciprocating compressors currently available for, and used in, commercial chillers, shows that the total relative residence time on these two adiabatic branches,

$$\Xi_{int} = \Xi_{comp} + \Xi_{exp}, \tag{7.8}$$

is of the order of magnitude 10^{-4} to 10^{-2} [Liang & Kuehn 1991; Bong *et al* 1990; Ng *et al* 1994]. The determination of these relative residence times from chiller construction follows in the next section. In the analysis that follows, we treat Ξ_{int} as a known fixed minimum value, and then optimize for the division of relative residence time and thermal inventory between the two heat exchangers.

In the governing energy and entropy balance equations, the heat exchanger control variables Ξ and $(mCE)'$ always appear as a product, rather than individually. Hence in the optimization exercise that requires

$$\partial \text{COP}/\partial \Xi_{evap} = 0 \quad \text{and} \tag{7.9}$$

$$\partial \text{COP}/\partial (mCE)'_{evap} = 0 \tag{7.10}$$

the solution must satisfy

$$\frac{\Xi_{evap}^{opt}}{1-\Xi_{int}} = \frac{(mCE)'^{opt}_{evap}}{C_0} \tag{7.11}$$

and hence

$$\frac{\Xi_{cond}^{opt}}{1-\Xi_{int}} = \frac{(mCE)'^{opt}_{cond}}{C_0} \tag{7.12}$$

where the additional "opt" superscript denotes values that maximize COP. Equations (7.11)–(7.12) are valid for all operational cooling rates. The global maximum for COP is obtained by additionally requiring

$$\partial \text{COP}/\partial Q_{evap} = 0. \tag{7.13}$$

Using a chiller's characteristic parameter values, and solving the 3 simultaneous equations (7.9), (7.10) and (7.13), one obtains $(mCE)'^{opt}_{evap}$ (defined as M_0 for economy of presentation in what follows) as the sole physically-admissible root of a sixth-order equation [Gordon *et al* 1997]. With the solution for $(mCE)'^{opt}_{evap}$, Ξ^{opt}_{evap} follows from Equation (7.11). The cooling rate at maximum COP is then

$$Q^{opt}_{evap} = T^{in}_{evap} M_0 (1 - \Xi_{int}) \times \frac{\left(C_0^3 - 3C_0^2 M_0 + 3C_0 M_0^2 - 2M_0^3\right)(1 - \Xi_{int}) - \Delta S_{int}\,\Xi_{int} C_0 \left(C_0 + 2M_0\right)}{2C_0^2\left\{(1 - \Xi_{int})M_0 - \Xi_{int}\Delta S_{int}\right\}}. \quad (7.14)$$

All the parameters in Equation (7.14) are known either from component specifications or from imposed chiller operating conditions.

D. COMPARISON WITH CHILLER EXPERIMENTAL DATA

The kind of detailed experimental data required for the evaluation of the equations derived above, and hence for determining to what degree commercial chillers have evolved to nominally optimal configurations, is not commonly available in the professional literature. To the best of our knowledge, there has been no validation of finite-time thermodynamic chiller models against actual experimental data for predicting the optimal way of allocating finite time and finite heat exchanger resources. We used the measurements reported in Chapter 6 for the nominal 10.7 kW and 7.56 kW water-cooled reciprocating chillers. Chiller parameters are summarized in Table 7.1. (The cooling rate is 10.7 kW here rather than the 10.5 kW noted in Chapter 6 because a different set of coolant temperatures is chosen for the chiller rating).

In estimating the relative residence times Ξ_{evap}, Ξ_{cond} and Ξ_{int}, we have assumed that Equations (7.11)–(7.12) apply to the two chillers analyzed here. The reason is that both chillers have the same heat exchanger construction (concentric tubes) and utilize the same coolant (water). Such a configuration should result in a chiller design that closely approaches the optimal relations dictated by Equations (7.11)–(7.12). Once one accepts that particular relation between the relative time in the heat exchangers and the heat exchanger $(mCE)'$ value, the values of Ξ_{evap}, Ξ_{cond} and Ξ_{int} emerge as solutions to the governing energy and entropy balance equations noted in Section B. For the nominal 10.7 kW chiller, Ξ_{int} turns out to be 2×10^{-2}, whereas for the nominal 7.56 kW chiller Ξ_{int} is 6.56×10^{-4}. Table 7.2 lists the principal comparisons

Table 7.1: Summary of experimental and computed values of the principal operating variables for two commercial water-cooled reciprocating chillers

chiller variable	chiller 1	chiller 2#
T^{in}_{cond}(K)	302.58 ± 0.05	282.48
T^{in}_{evap} (K)	285.54 ± 0.05	282.43
T_{cond} (K)	320.0 ± 0.9	302.91
T_{evap} (K)	267 ± 2	266.09
Q_{evap} (kW)	10.7 ± 0.2	7.56
Q^{leak}_{cond} (kW)	0.006 ± 0.4	0.15
Q^{leak}_{evap}(kW)	0.219 ± 0.3	0.14
Q^{leak}_{comp}(kW)	0.127 ± 0.01	0.08
$(mCE)'_{cond}$ (kW K^{-1})§	1.53 ± 0.08	0.927
$(mCE)'_{evap}$ (kW K^{-1})§	1.31 ± 0.11	0.926
ΔS_{int} (kW K^{-1})	0.00511 ± 0.0003	0.00279
$\Xi_{int} = \Xi_{comp} + \Xi_{exp}$	0.02 (*)	0.000656
1/COP	0.373 ± 0.006	0.263
1/COP excluding heat leaks	0.35 ± 0.01	0.250

Experimental uncertainties unavailable from [Liang & Kuehn 1991].
*The value of Ξ_{int} is deduced from a combination of several experimental measurements and from refrigerant data from thermodynamic tables.
§The *mCE* heat exchanger values cited in Chapter 6 differ from the $(mCE)'$ values used here by the relative residence time factor Ξ.

between: (1) predicted optimal operating conditions and predicted performance, versus (2) actual chiller performance.

For the experimental cases analyzed here, the dissipation happens to be divided approximately equally between external irreversibilities in the heat exchangers, and internal losses in the compressor and throttler. The internal dissipation is dominated by the compressor. The instantaneous rate of entropy production during throttling is highest among all chiller components. But the total dissipation during expansion is negligibly small because of the near-sonic speeds, and hence vanishingly small relative residence time, at which the refrigerant flows during expansion. The fact that chiller manufacturers have not invested in improvements of expansion devices (such as small power-recovery

Table 7.2 Comparison of theoretical optimization predictions and experimental performance data for two commercial reciprocating chillers

chiller variable	nominal 10.7 kW chiller	nominal 7.56 kW chiller#
δS_{comp} Ξ_{comp} (kW K^{-1})	0.0047 ± 0.0001	0.003
δS_{exp} Ξ_{exp} (kW K^{-1})	0.0004 ± 0.0002	0.0003
predicted $(mCE)'^{opt}_{evap}$ (kW K^{-1})	1.364	0.892
actual $(mCE)'_{evap}$ (kW K^{-1})	1.31 ± 0.11	0.926
predicted Ξ^{opt}_{evap} (kW)	0.4700	0.4814
actual Ξ_{evap}	0.45 ± 0.01	0.4998
predicted Q^{opt}_{evap} (kW)	10.66	6.426
actual Q_{evap} (kW)	10.7 ± 0.2	7.56
predicted minimum value of 1/COP	0.350	0.246
actual 1/COP	0.373 ± 0.006	0.263

#Experimental uncertainties unavailable from [Liang & Kuehn 1991].

turbines) is probably reflected by the small potential gain. For the heat exchangers, the optimal strategy is to spend more time where there is greater thermal inventory: the optimal relative residence time and the optimal $(mCE)'$ values are proportional to one another. The optimization for relative residence time in the compressor and throttler is uncomplicated because COP decreases monotonically as relative residence times increase. The comparison between chiller performance at nominally optimal configurations and actual chiller performance attests to the fact that, perhaps without even being cognizant of the extra degree of freedom of relative residence time inherent at the early stages of chiller design, manufacturers have evolved commercial units that are impressively close to the theoretical thermodynamic optimum.

E. EQUIVALENCE OF MAXIMIZING COP AND MINIMIZING UNIVERSAL ENTROPY PRODUCTION

By now, it's clear that minimizing entropy production in the chiller is equivalent to maximizing the chiller's COP. The chiller's thermal interactions with its reservoirs also produce entropy in those environments. More global analyses of entropy production often address the entropy production in the universe, *i.e.*, in the chiller and its surroundings. The objectives of maximizing COP and minimizing entropy production in the universe are not necessarily the same. For certain specific sets of constraints, however, it is possible for the strategies of maximizing COP and minimizing dissipation in the universe to be identical. We will now show that for constraints of practical interest to chiller manufacturers and designers these two objectives can turn out to be equivalent. This could be more than a passing esoteric observation because it means that, for the particular limitations within which chiller producers typically work, they would appear to be investing in designs that are globally minimally dissipative, as well as being maximally efficient. The total entropy production in the universe ΔS_u can be calculated from the equations listed in Section B:

$$\Delta S_u = (mC)'_{\text{cond}} \ln\left[1 + \frac{Q_{\text{evap}}}{(mC)'_{\text{cond}} T^{\text{in}}_{\text{cond}}}\left\{1 + \frac{1}{\text{COP}} - \frac{Q^{\text{leak}}_{\text{net}}}{Q_{\text{evap}}}\right\}\right] + (mC)'_{\text{evap}} \ln\left[1 - \frac{Q_{\text{evap}}}{(mC)'_{\text{evap}} T^{\text{in}}_{\text{evap}}}\right] + \frac{Q^{\text{leak}}_{\text{net}}}{T_{\text{env}}} \tag{7.15}$$

where for brevity of notation we define

$$Q^{\text{leak}}_{\text{net}} = Q^{\text{leak}}_{\text{cond}} + Q^{\text{leak}}_{\text{comp}} - Q^{\text{leak}}_{\text{evap}} \tag{7.16}$$

where T_{env} is temperature of the chiller environment. Equation (7.15) highlights the increase in ΔS_u due to finite heat reservoirs and due to the monotonic variation of COP with respect to $(mC)'_{\text{cond}}$ and $(mC)'_{\text{evap}}$. In the limit of infinite reservoirs, Equation (7.15) reduces to

$$\Delta S_u = Q_{\text{evap}} \left[\frac{1 + \dfrac{1}{\text{COP}} - \dfrac{Q_{\text{net}}^{\text{leak}}}{Q_{\text{evap}}}}{T_{\text{cond}}^{\text{in}}} - \frac{1}{T_{\text{evap}}^{\text{in}}} \right] + \frac{Q_{\text{net}}^{\text{leak}}}{T_{\text{env}}}. \tag{7.17}$$

For fixed reservoir temperatures $T^{\text{in}}_{\text{cond}}$ and $T^{\text{in}}_{\text{evap}}$, maximizing COP at each value of Q_{evap} is hence equivalent to minimizing ΔS_u (provided the heat leak term in Equation (7.15) is approximately constant or negligibly small). In asking under what chiller operating conditions dissipation is minimal, one usually considers Q_{evap} as the control variable, and hence reaches the obvious conclusion that only in the limit of $Q_{\text{evap}} \to 0$ will ΔS_u vanish and hence be minimized. An alternative conventional view is to treat the reservoir (coolant) temperatures as control variables, in which case similarly simple results are realized. In considering practical chiller design and operation, the germane query is how to optimize a chiller characteristic at a given cooling rate and at fixed reservoir temperatures. But if the reservoir temperatures and Q_{evap} are fixed, what are the control variables? In this optimization study, the design degrees of freedom are the relative residence times Ξ, and the division of total $(mCE)'$ between the condenser and evaporator. They are very different control variables than cooling rate and coolant temperatures, but they are the controls that chiller manufacturers have at their disposal and, wittingly or unwittingly, select when they configure commercial devices. Inspection of Equation (7.15) then reveals that, with the allocation of the finite available resources serving as the controls, the aims of maximum COP and minimum ΔS_u turn out to be the same.

F. CLOSURE

This chapter represents a sort of thought experiment. No one installing or using a cooling system would deem the exercise of interest or importance. Chiller manufacturers may not have consciously contemplated the extra degrees of freedom, and hence optimization, implied in our approach. Yet the optimization variables are physically meaningful, even if they have not been the meaningful variables incorporated in intentional manufacturer optimizations performed in the past. The application of finite-time thermodynamics to chiller design - at a very early stage of the design process where selected properties of each component can still be viewed as open-ended - offers additional room for improvement in thermodynamic performance. An intriguing finding here is that, even without knowingly incorporating a finite-time optimization into their designs, chiller manufacturers appear to have produced commercial devices that are impressively close to the theoretical optimum.

Chapter 8

COOLANT FLOW RATE AS A CONTROL VARIABLE

"A theory may be so rich in descriptive possibilities that it can be made to fit any data." - Phillip Johnson-Laird, The Computer and the Mind

A. BACKGROUND TO THE PROBLEM

Although most commercial chillers are designed and installed to operate at constant coolant flow rates, the power consumption of their compressors can be altered by modifying the design to incorporate variable coolant flow rates. The issue addressed in this chapter is how to model that explicit flow-rate dependence within the analytic chiller model of the preceding chapters. We'll show how the model is easily expanded to account explicitly for the influence of coolant flow rate. Then we'll validate model predictions against an extensive set of experimental measurements from a large commercial centrifugal chiller [Gordon *et al* 2000].

Coolant flow rate, in particular at the condenser, could be an additional control variable. Although one incurs the complication and expense of a variable-speed coolant pump, chiller efficiency can be improved. For example, in the centrifugal chiller reported below, the power consumption of the compressor at a given cooling rate and fixed coolant temperatures can change by as much as 10% over a range of realistically implementable condenser coolant flow rates.

Before embarking upon model development, we should address the perceived value of this type of information for the chiller installer and consumer. The simple chiller model adopted here accounts for the power consumption of the compressor only. This also accounts for pressure drops on the refrigerant side of the heat exchangers, but does not relate to pressure drops on the coolant side.

Now in general, the designer is interested in the minimization of the total electrical power consumption during part-load conditions, which depends on pressure drops on both sides of the heat exchangers. During part-load, when condenser heat rejection lessens, lowering the condenser flow rate permits the pumping power to be diminished, provided the required coolant inlet and outlet temperatures remain roughly unchanged. Lower condenser flow rates also reduce erosion within the heat exchanger tubes. However, lower condenser flow rate would increase compressor power.

Alternatively, the compressor power can be reduced at fixed coolant pumping power. The minimum *total* electrical power consumption will be the lesser of the two strategies. So under certain conditions, lowering the condenser coolant flow rate can result in a greater reduction in the total power usage than the strategy of reducing compressor power via a lowering of the mean coolant temperature across the condenser.

In this chapter, we will be focusing upon adapting our analytic chiller model – which was developed for *constant* coolant flow rates – to account for *variable* coolant flow rate. The model addresses the *compressor* power consumption (as is common practice in reporting chiller efficiency), and *not* the power consumption of the coolant pumps. Hence the ultimate minimization of the *total* power consumption of a particular system should combine the model results developed below with coolant pump power consumption data.

The savings in *liquid* pumping power, as a percentage of the total chiller plant electrical consumption, is usually small. With the availability of computer-based chiller controls, however, the additional task of saving on pumping power is easily achieved through software modifications.

Why is it practical to vary condenser, but not evaporator, coolant flow rate? In the water-cooled condenser, moisture is shed to the environment at the cooling tower, whereas in the cooling coils, moisture is deposited on the tubes and fins. The dehumidification process of the cold air within the cooling coils limits the range of evaporator coolant flow rate. Reducing evaporator coolant flow rate amounts to raising the effective surface temperature of the cooling coils, which in turn lessens dehumidification. Namely, although sensible heat can still be removed at a sufficient rate, the same does not necessarily pertain to latent heat removal. The same type of potential heat-removal bottleneck does not occur in the condenser.

The question then arises whether one can accurately correlate chiller performance (specifically, *compressor* power consumption) with coolant flow rate. That information could be incorporated into a simple control strategy for minimizing total power consumption at prescribed operating conditions. A fringe benefit would be an analytic diagnostic tool with which future changes in chiller performance could be detected and quantified.

In fact, this chapter was motivated by precisely this problem having been raised by a firm commissioned to develop a nominally optimal chiller operating strategy for a large centrifugal chiller used to cool clean rooms in the semiconductor-processing industry. Extensive experimental measurements of how chiller efficiency and cooling rate vary with condenser coolant flow rate were provided. The challenge was to model the dependence of chiller performance on condenser coolant flow rate in a simple analytic algorithm. The company intended to use that algorithm to vary pump speed and thereby to minimize total power consumption.

One can invoke the thermodynamic chiller model developed in earlier chapters, and try to introduce coolant flow rate in a physically meaningful fashion. Because the model parameters correspond to clearly identifiable physical characteristics, one can predict the explicit dependence of the model parameters on coolant flow rate. Employing standard regression procedures on the measured data set, one can then characterize the chiller in terms of several parameters, one of which would be uniquely linked to the influence of condenser coolant flow rate.

We will test the ability of the model to correlate centrifugal chiller performance with the extensive experimental measurements provided by the chiller manufacturer. The root-mean-square (rms) error of model predictions turns out to be less than the estimated measurement error. In addition, we'll demonstrate that the chiller's characteristic parameters can equally well be determined from a relatively small number of judiciously-selected measurements. This reinforces our earlier claims that with the analytic modeling procedure, extensive data sets are often not required.

As an additional test of model predictions, we will determine the chiller characteristic parameters based upon a censored data set that excludes operating points of extreme coolant flow rate and extreme coolant temperatures. Chiller performance at the extreme operating conditions is then predicted. The rms error of the model predictions remains unaltered, hence confirming the value of the model in extrapolating to operating conditions beyond measured values (but within a reasonable operating range). We'll apply additional self-consistency checks to confirm that the statistical accuracy of the analytic thermodynamic model also preserves the physical characteristics of the model parameters.

B. ADAPTING THE ANALYTIC CHILLER MODEL TO INCORPORATE COOLANT FLOW RATES

Recall the basic chiller performance formula derived in Chapter 5, where 3 parameters characterize chiller behavior: (1) the effective rate of internal entropy production ΔS_{int} over various operating conditions, (2) the effective overall thermal resistance R of the heat exchangers, and (3) the equivalent heat leak Q_{eqv}^{leak} between the refrigerant and its environments:

$$\frac{T_{evap}^{in}}{T_{cond}^{in}}\left[1+\frac{1}{COP}\right]=1+\frac{T_{evap}^{in}\Delta S_{int}}{Q_{evap}}+\frac{Q_{eqv}^{leak}}{Q_{evap}}\left[\frac{T_{cond}^{in}-T_{evap}^{in}}{\mathrm{T}_{cond}^{in}}\right]+\frac{Q_{evap}R}{T_{cond}^{in}}\left[1+\frac{1}{COP}\right]. \tag{8.1}$$

To characterize the chiller quantitatively, one typically measures COP and Q_{evap} at assorted values of coolant inlet temperatures. Then, with a multiple linear-regression procedure, one calculates the 3 chiller parameters ΔS_{int}, Q_{eqv}^{leak} and R from Equation (8.1). At the outset, we stress that this analysis relates to steady-state performance only.

In this case study, we are examining a centrifugal, as opposed to a reciprocating, chiller. There is no assurance that, in centrifugal chillers, the purely statistical best-fit values of these 3 model parameters will correspond to the actual values of the physical parameters they are purported to represent. The experimental studies we reported in Chapter 6 for vapor-compression cycles offered verification of the diagnostic and predictive power of the thermodynamic model represented in Equation (8.1). Specifically, the internal entropy production, heat leaks and heat exchanger conductances were determined independently with intrusive measurements, and compared against the regressed values of the 3 parameters determined from Equation (8.1) with common non-intrusive measurements, with excellent agreement.

Similar verification experiments on centrifugal chillers are problematic and were not attempted. The difficulty stems from the typical large size (cooling capacity) of commercial centrifugal machines. In addition, the equipment warranty (and hence the cost of the associated repair) militates against tempering the state of the principal chiller components. Based on the confidence gained from the successful application of the model to reciprocating chillers, and considering the common vapor-compression principle that centrifugal compressors share with their reciprocating counterparts, it seemed reasonable to apply the same model to centrifugal chillers.

The overall thermal resistance R in Equation (8.1) can be expressed in terms of the coolant volumetric flow rate V, coolant specific heat C, coolant density ρ, and heat exchanger effectiveness values at the condenser and evaporator

$$R = R_{\text{cond}} + R_{\text{evap}} = \frac{1}{(V\rho CE)_{\text{cond}}} + \frac{1}{(V\rho CE)_{\text{evap}}} \tag{8.2}$$

when coolant inlet temperatures are used in Equation (8.1). It turned out that the manufacturer's data in this study were reported in terms of condenser inlet and evaporator outlet temperatures [Trane 1996]. When a coolant outlet temperature is to be used in Equation (8.1), the heat exchanger effectiveness E in Equation (8.2) must be modified to $\frac{E}{1-E}$. In this specific case, then, Equation (8.2) becomes

$$R = \frac{1}{(V\rho CE)_{\text{cond}}} + \frac{1-E_{\text{evap}}}{(V\rho CE)_{\text{evap}}}. \tag{8.3}$$

Although the regression of Equation (8.1) is only weakly sensitive to heat leaks, the heat leak term is retained in the model for maximum accuracy and for preserving the complete physical picture of the chiller.

C. EXPLICIT ACCOUNTING FOR THE INFLUENCE OF COOLANT FLOW RATE

The problem posed here is how to explicitly incorporate the role of condenser flow rate. The qualitative trend is clear at the outset. Increasing coolant flow rate lowers heat exchanger thermal resistance, which in turn raises COP. The challenge is to quantify the observation in a readily-implemented analytic model. We stress that the variables being modeled are cooling rate and COP, and do not include consideration of the power required for coolant pumps. A water-cooled centrifugal chiller with tube-in-shell heat exchangers was under consideration [Trane 1996]. Although all three chiller parameters in Equation (8.1) can in principle change with coolant flow rate, for realistic operating conditions the only significant dependence should arise from the heat exchanger thermal resistance. This point is reinforced from an analysis of the experimental data reported below. With the refrigerant in the condenser being characterized by an effective isothermal (process-average) temperature, the heat exchanger effectiveness can be expressed as

$$E = 1 - \exp\left[\frac{-UA}{V\rho C}\right] \tag{8.4}$$

where U and A are the heat exchanger's thermal conductance and heat-exchange area, respectively. For turbulent flow, the thermal conductance U is approximately proportional to $V^{0.8}$ [Mills 1992]. Hence E can be written as

$$E = 1 - \exp\left[\frac{-\kappa}{V^{0.2}}\right] \tag{8.5}$$

where κ is a positive constant that characterizes the particular heat exchanger. The ruling thermal resistance is that of sensible heat transfer on the coolant side of the heat exchanger (the refrigerant side benefits from latent heat transfer and hence a far smaller thermal resistance).

For the system analyzed here, with fixed V_{evap} and variable V_{cond}, we can now modify Equation (8.1) to

$$\frac{T_{\text{evap}}^{\text{out}}}{T_{\text{cond}}^{\text{in}}}\left[1+\frac{1}{\text{COP}}\right] = 1 + \frac{T_{\text{evap}}^{\text{out}}\Delta S_{\text{int}}}{Q_{\text{evap}}} + \frac{Q_{\text{eqv}}^{\text{leak}}}{Q_{\text{evap}}}\left[\frac{T_{\text{cond}}^{\text{in}} - T_{\text{evap}}^{\text{out}}}{T_{\text{cond}}^{\text{in}}}\right]$$
$$+\frac{Q_{\text{evap}}}{T_{\text{cond}}^{\text{in}}}\left[1+\frac{1}{\text{COP}}\right]\left[\frac{1}{\left\{V\rho C\left[1-\exp\left\{\frac{-\kappa}{V^{0.2}}\right\}\right]\right\}_{\text{cond}}} + R'_{\text{evap}}\right] \tag{8.6}$$

where

$$R'_{\text{evap}} = \frac{1 - E_{\text{evap}}}{(V\rho CE)_{\text{evap}}}. \tag{8.7}$$

In Equation (8.6), the chiller is characterized by four parameters (ΔS_{int}, $Q_{\text{eqv}}^{\text{leak}}$, κ and R'_{evap}), and the explicit influence of V_{cond} is accounted for.

Via non-linear regression, one can calculate these four parameters from ordinary measurements of COP as a function of cooling rate and coolant temperatures, but performed at several values of coolant flow rate. For the centrifugal chiller analyzed below, the coefficient κ turned out to be statistically insignificant. Namely, κ was sufficiently large that COP is insensitive to it. Hence the condenser heat exchanger effectiveness can be regarded as sufficiently close to unity (*i.e.*, large κ) that the term $\left\{V\rho C\left[1-\exp\left[\frac{-\kappa}{V^{0.2}}\right]\right]\right\}_{\text{cond}}$ in Equation (8.6) can be approximated as $(V\rho C)_{\text{cond}}$. In this instance, Equation (8.6) reverts to a three-parameter model (ΔS_{int}, $Q_{\text{eqv}}^{\text{leak}}$ and R'_{evap}) where the parameters can be ascertained with multiple linear regression, and one emerges with a simple and explicit prediction of how chiller performance varies with condenser coolant flow rate.

D. EXPERIMENTAL DETAILS

A water-cooled centrifugal chiller was considered, with a nominal rated cooling rate of 2464 kW [Trane 1996]. At rated conditions, the coolant volumetric flow rates are $V_{\text{evap}} = 81.5$ l s^{-1}, $V_{\text{cond}} = 133$ l s^{-1}, and the coolant temperatures in °C are: $T_{\text{evap}}^{\text{in}} = 12.8$, $T_{\text{evap}}^{\text{out}} = 6.7$, $T_{\text{cond}}^{\text{in}} = 23.9$ and $T_{\text{cond}}^{\text{out}} = 28.9$. The refrigerant is R-123, with a refrigerant charge of 750 kg.

The manufacturer provided experimental measurements of COP and Q_{evap} at the rated conditions of $T_{\text{cond}}^{\text{in}} = 23.9$°C and $V_{\text{evap}} = 88$ l s^{-1}. A total of 579 different operating conditions were reported, that included 9 different values of V_{cond} from 93 to 160 l s^{-1}. The data covered 7 values of $T_{\text{evap}}^{\text{out}}$ from 4.4 to 8.7°C, and 11 different cooling rates from 370 to 2464 kW. COPs ranged from 4.5 to 8.7. The estimated experimental uncertainty in COP values was 5%.

The chiller's characteristic performance curves, graphed in Figure 8.1 (each curve comprising 10 measured points), illustrate the influence of condenser coolant flow rate, and reveal basic anticipated trends. For example, from previous analyses we know that the approximately linear regime at lower cooling rates is dominated by internal dissipation, while external or finite-rate heat exchange losses grow significant in the non-linear region at higher cooling rates. Hence the fact that all the curves in Figure 8.1 are roughly linear with the same slope at lower cooling rates attests to the rate of internal dissipation being insensitive to condenser coolant flow rate. The displacement among the curves, and the growing differences at higher cooling rates, indicate that altering condenser coolant flow rate manifests itself in modifying external losses, as would be expected from Equation (8.6).

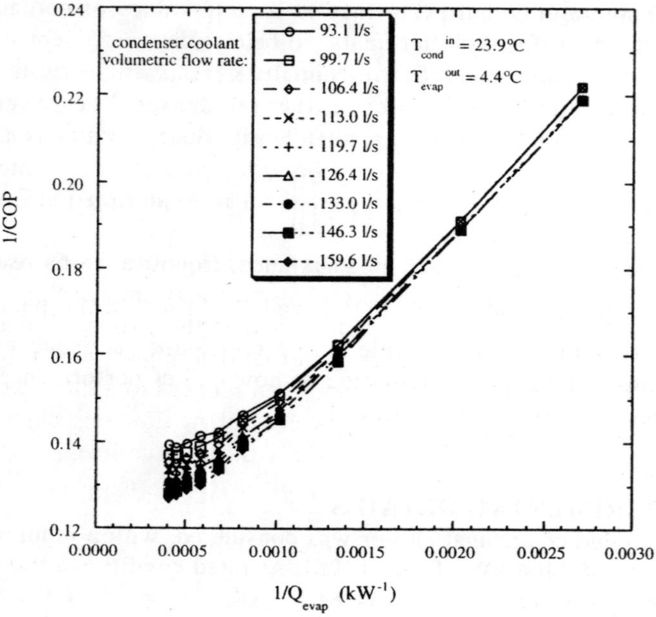

Figure 8.1: Characteristic chiller plot of 1/COP against 1/(cooling rate) at 9 values of condenser coolant volumetric flow rate.

What underlies the improvement in COP as condenser coolant flow rate is raised? Primarily, higher coolant flow rate lowers the thermal lift in the compressor (for a given cooling rate). The lower thermal lift in turn reduces the required compressor power, hence raising COP. There is a competing but smaller effect of greater compressor work being required due to the heightened pressure drop of the refrigerant (again, at a particular cooling rate). For a given increase in condenser coolant flow rate, the associated change in the compressor's thermal lift is more pronounced at higher cooling rates. The greater impact of condenser coolant flow rate at higher cooling rates is evident from the data plotted in Figure 8.1. When condenser flow rate spans a factor of 1.7, the improvement in COP varies from 1% at low part-load conditions to 10% at full load. These figures, and the principal features of Figure 8.1, pertain to one particular set of fixed coolant temperatures. However, they remain the same for other pairs of coolant temperatures within the range covered by the experiments.

E. APPLICATION OF THE MODEL AND EXPERIMENTAL CONFIRMATION

First, a non-linear regression is performed on Equation (8.6). An equivalent mathematical procedure (to check the non-linear routine) is also conducted whereby the value of κ in Equation (8.6) is fixed, a multiple linear regression for the other three chiller parameters is performed, and the exercise is repeated for a broad range of κ values. The two methods yield the same results.

It turned out that this particular chiller had a condenser heat exchanger with an effectiveness of essentially unity. Namely, the regressed value of κ was large enough that the exponential term in Equation (8.6) is negligible; hence the exact determination of κ becomes statistically meaningless. The problem then reduces to a three-parameter model, with a simple explicit condenser coolant flow rate dependence, where the parameters can be extracted with multiple linear regression. Using all 579 data points with Equation (8.6) in the preliminary regression, we obtained

$$\Delta S_{int} = 0.216 \text{ kW K}^{-1},\ R'_{evap} = 0.00226 \text{ K kW}^{-1} \text{ and } Q_{eqv}^{leak} = -121 \text{ kW} . \quad (8.8)$$

(Note that the effective heat leak term in Equation (8.6) constitutes only several percent of the total. In addition, Q_{eqv}^{leak} can be negative because the heat leak from the environment into the evaporator can

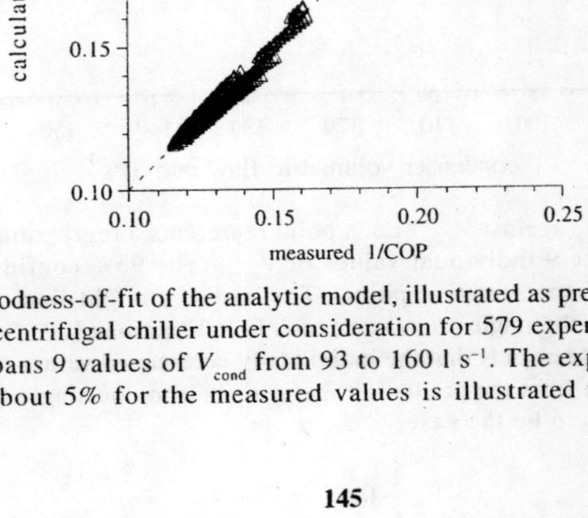

Figure 8.2: Goodness-of-fit of the analytic model, illustrated as predicted versus measured 1/COP for the centrifugal chiller under consideration for 579 experimental measurements. The data set spans 9 values of V_{cond} from 93 to 160 l s^{-1}. The experimental uncertainty in 1/COP of about 5% for the measured values is illustrated for one representative point.

exceed the heat leak from the condenser to ambient.) When the parameters in (8.8) are used to correlate chiller COP for the same data set, the rms error is 2%, which is well below experimental error. The maximum error in correlating any measured COP value was only 4.3%. Figure 8.2 displays these results graphically.

Three other checks can be performed on the data set. In one, in order to ascertain the model's predictive capabilities, we perform the multiple linear regression for the chiller parameters by separately omitting the following extreme situations from the data set: (*a*) the highest V_{cond}; (*b*) the lowest V_{cond}; (*c*) the highest T_{evap}^{out}; or (*d*) the lowest T_{evap}^{out}. Each case represents about 10% of the data set. The regressed parameters are then used in Equation (8.6) to predict chiller COP for the excluded extreme operating conditions. The rms error for the predictions of these isolated extreme points turns out to differ negligibly from the overall rms error of 2%.

The second check engenders performing the regression analysis at each of the 9 individual values of V_{cond} with Equation (8.1) (with around 64 data points at each flow rate). Two important results should emerge. One is that ΔS_{int} should not vary significantly from one flow rate to the next. That this is indeed the case can be seen from Figure 8.3.

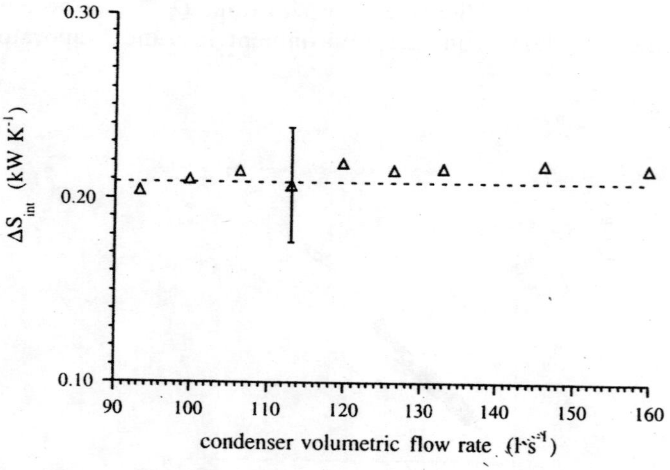

Figure 8.3: Plot of ΔS_{int} against $V_{cond.}$ Each point represents a regression fit performed on Equation (8.1) at the 9 individual values of V_{cond}. The 95% confidence intervals are illustrated for one representative point. The broken straight line corresponds to the prediction made with the single grand regression fit to Equation (8.6). Self-consistency in the model predictions demands that the individually-determined values (plotted points) not deviate significantly from the prediction that derives from the single fit to the entire data set, which appears to be the case.

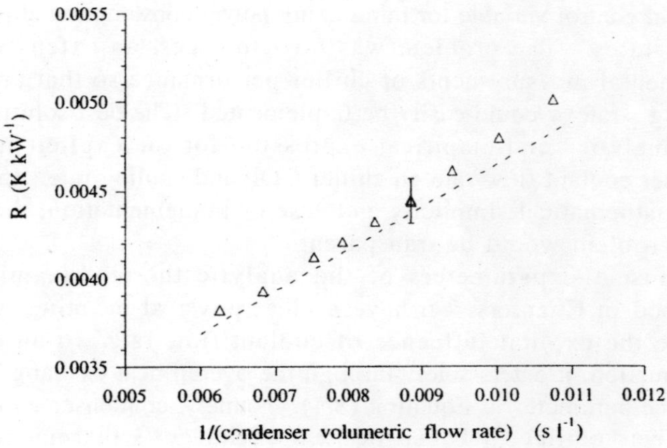

Figure 8.4: Plot of R against $1/V_{cond}$. Each of the points represents a regression fit performed on Equation (8.1) at the 9 separate values of $V_{cond.}$ The 95% confidence intervals are illustrated for one representative point. The broken straight line corresponds to the prediction made with the single grand regression fit to Equation (8.6). Self-consistency in the model predictions demands that the individually-determined values (plotted points) not deviate significantly from the prediction that derives from the single fit to the entire data set. This is indeed confirmed.

The other is that R should be linear in $1/V_{cond}$ (Equations (8.2) and (8.3) with $E_{cond} \approx 1$). Furthermore, that linear relation should be approximately the same as the result from a single grand regression fit to the entire data set with Equation (8.6). Confirmation of this check is illustrated in Figure 8.4

The third check is to confirm that the model does not need a large data bank to provide an accurate correlation for chiller performance. We gradually reduced the number of data sets considered in the multiple linear regression from the original value of 579 to as little as 24. In all cases, the 95% confidence intervals for the 3 coefficients are always within those obtained when the original 579 data sets were used. Furthermore, in all cases, the rms error of correlation is still within the experimental error of 5%. However, when the number of data sets is reduced to the level of a few dozen, there are a handful of data points where the error exceeds 5%.

F. CLOSURE

The project described in this chapter originated as part of an optimization exercise for a large installed centrifugal chiller, where the ability to vary the condenser coolant flow rate was to be introduced as an additional control variable for minimizing power consumption at prescribed cooling rates. The problem was how to digest an extensive set of experimental measurements of chiller performance so that an optimal operating strategy could easily be implemented. The best solution would be an analytic semi-empirical expression for the explicit impact of condenser coolant flow rate on chiller COP and cooling rate. In addition to the mathematical simplicity and ease of implementation, the physics of the problem would be transparent.

Because the parameters of the analytic thermodynamic model developed in Chapters 4–6 have a clear physical meaning, we could evaluate the explicit influence of coolant flow rate. To an excellent approximation, it enters solely through the overall heat exchanger thermal resistance parameter in Equation (8.1). Namely, condenser coolant flow rate affects chiller COP via the heat exchanger's thermal resistance in principle in two ways: (*a*) a simple explicit dependence on the volumetric flow rate; and (*b*) an implicit dependence via the heat exchanger's effectiveness. We found that the implicit dependence turns out to be statistically insignificant, so that the correlation procedure reduces to a relatively simple multiple linear regression fit. With a relatively simple regression analysis of experimental measurements of chiller performance at 9 different flow rates, one can verify the ability of the model to correlate and predict the performance data, as well as the dependence of chiller performance on condenser coolant flow rate. Several checks confirmed that the regressed model parameters are characteristic of the physical properties they are supposed to represent.

The central issue addressed in this chapter has been a genuine chiller control problem of concern to field engineers and monitors. We have provided an accurate means of correlating chiller COP when the coolant flow rate of the condenser can be varied significantly. We had the good fortune of being provided with an extensive set of experimental measurements against which to compare our predictions. One attractive aspect of the findings is the confirmation that chiller COP can be predicted over a large range of chiller performance variables from just a handful of measured data.

Chapter 9

OPTIMIZATION OF ABSORPTION SYSTEMS

"Have no fear of perfection - you'll never reach it." -
Salvador Dali

A. OBJECTIVES AND MOTIVATION

Let's adopt the perspective of the designer or manufacturer of absorption systems. The operating mode can be chiller, heat pump or heat transformer. The device is still at the design and assembly stage – not yet operational. We can modify any or all of the 4 heat reservoirs, as well as the *mCE* values of the heat exchangers.

For chillers and heat pumps, each possible configuration corresponds to a particular division of total heat rejection between the condenser and the absorber. That is why we have treated this partition as a meaningful control variable. Similarly for heat transformers, the division of the total heat input between the generator and the evaporator is regarded as the corresponding control variable. For all operating modes, one particular value of this control variable will maximize COP. So determining that value is of considerable worth to the designer and manufacturer.

The spirit of our analytic, simplified, approximate modeling schemes is similar to that stated earlier for mechanical chillers. Namely, we are not supplanting massive simulations wherein every property of every component is analyzed in great detail, and every possible chiller performance variable is generated. Rather, we are attempting to capture the key performance features and trends of absorption machines, *e.g.*, the points of maximum COP and maximum useful effect, the variation of COP with useful effect, and how the machine's characteristic performance curve deviates from endoreversible behavior. In this sense, we are again establishing a sort of "base case" analysis with which manufacturers and designers can obtain acceptably accurate predictions of optimal configurations.

As we've seen in Chapters 4 and 5, COP varies significantly with useful effect. Because of the competition between external and internal losses, COP is maximized at a particular value of useful effect. This constitutes another local optimization. A more global optimization involves maximizing COP as a function of both useful effect and the partition of heat rejection (or heat input) at two reservoirs.

There will also be a point of maximum useful effect (recall Figure 5.1) – an inherent bound imposed by the irreversibilities in the generator. Since absorption machines are often driven by inexpensive low-grade heat, identifying the point of maximum useful effect can be of importance to manufacturers and designers.

Once we find the theoretical (but realistic) optimum operating conditions, we can establish the maximum possible improvement in device COP for a given technology, that is, for a fixed quality of the individual components. In addition, we can examine to what extent commercial absorption systems have evolved to optimal or near-optimal designs. In cases where a noticeable discrepancy exists, the model enables us to identify and quantify the performance bottlenecks. In these senses, these exercises also offer a diagnostic capability.

The analytic model derived in Chapters 4 and 5 enables us to answer all of the above issues, provided we know the parameters in the governing equation. As we observed in Chapter 5, this is not feasible if we are restricted to typical catalog data. However, with the perspective of the manufacturer assembling the system, we can measure these parameters. With this superior level of characterization of the absorption device, we proceed to examine these problems.

Finally, as sketched in Chapter 2, some higher-efficiency absorption units are designed and built with enhanced regenerative heat exchange. They are usually referred to as double-stage or triple-stage units, depending on the number of heat exchangers and generators introduced. The regenerative heat exchange can also be performed in series or in parallel. Only the high-temperature generator, condenser, evaporator and absorber are in thermal communication with the heat reservoirs. We treat the heat transfer irreversibilities at these heat reservoirs as external losses. When heat transfer dissipation occurs within the regenerative scheme, it will be assimilated as part of the internal losses. This categorization of irreversibilities permits us to treat single-stage, double-stage and triple-stage units within one united framework.

B. EXPERIMENTAL DATA, COMPUTER SIMULATION RESULTS AND DEVICE OPTIMIZATION

B1. The devices studied

The type of detailed data required for these optimization studies is not provided by manufacturer catalogs or most journal papers. The data may be determined by manufacturers in-house in developmental studies, but do not commonly find their way into the professional literature. Nonetheless, we found one manufacturer report [Carrier 1962] and several computer simulation studies [Chuang & Ishida 1990; Abrahamsson *et al* 1995; Zhuo & Machielsen 1996] that suffice for these analyses. The absorption machines they studied, and which we'll be examining here are:

(1) two single-stage chillers;
(2) a double-stage series-cycle chiller;
(3) a double-stage parallel-cycle chiller;
(4) a single-stage heat pump;
(5) two single-stage heat transformers;
(6) a double-stage heat transformer; and
(7) two triple-stage heat transformers (one of which is a special design which compromises heat exchanger inventory in favor of compactness).

The principal device characteristics and performance variables are summarized below in Tables 9.1–9.5. The experimental procedures for determining all the requisite chiller parameters were reviewed in Chapter 3.

B2. Comparison of device performance and predicted optima

First, we confirmed that system performance variables predicted by the model from component input parameters agreed with the experimental measurements and simulation results. Then we proceeded to calculate the optimal operating points for each system. These calculated optima are listed in Tables 9.2 and 9.5 for comparison against measured and simulated operating characteristics. Figures 9.1–9.4 are sample characteristic plots of 1/COP against 1/(useful effect). It turns out that the optimal operating region is broad, which means that system efficiency in that regime is tolerant to variations in useful effect, heat exchanger allocation or design modifications.

As an illustration of the shortcomings of the endoreversible chiller model, we have also plotted the corresponding endoreversible (*i.e.*, accounting for external losses only) curve in Figure 9.1. The failure of the endoreversible model to capture both quantitative and qualitative aspects of chiller behavior is apparent. We will return to the problems associated with endoreversible models for absorption machines in Chapter 11.

B3. Absorption chillers and heat pumps: diagnostics and design conclusions

For the heat pump and chillers, in 4 of the 5 units analyzed the prescribed configurations are close to the calculated optimal variables

Table 9.1: Absorption heat pump and chillers: summary of experimental data and simulated results. Figures from [Carrier 1962] are experimental data. Those from [Chuang & Ishida 1990] and [Abrahamsson *et al* 1995] are computer simulation results.

Variable ↓	single-stage chiller [Carrier 1962]	single-stage chiller [Chuang & Ishida 1990]	double-stage series cycle chiller [Chuang & Ishida 1990]	double-stage parallel-cycle chiller [Chuang & Ishida 1990]	single-stage heat pump [Abrahamsson *et al* 1995]
T_{gen}^{in} (°C)	117.6	170.7	170.7	170.7	149
T_{evap}^{in} (°C)	11.7	12.0	12.0	12.0	55
T_{abs}^{in} (°C)	29.4	32.0	32.0	32.0	90
Q_{gen}^{leak} (kW)	0	221	0	0	0
Q_{evap}^{leak} (kW)	15.3	0	131	131	0
$(UA)_{gen}$ (kW K^{-1})	63.4	814	460	408	282
$(mCE)_{cond}$ (kW K^{-1})	57.4	73.5	498	509	538*
$(mCE)_{evap}$ (kW K^{-1})	86.6	503	503	503	500*
$(mCE)_{abs}$ (kW K^{-1})	69.9	699	722	650	266*
$(mC)_{abs}$ (kW K^{-1})	146	1780	1370	1300	∞
ΔS_{int} (kW K^{-1})	0.202	1.30	1.39	1.04	0.113

*Because heat transfer is dominated by latent heat, rather than sensible heat, in these cases this is actually the *UA* value

(Figure 9.1 being an illustration). This is consistent with our findings in previous chapters for mechanical chillers that manufacturers have empirically developed units that operate near their theoretical maximum-COP points. Figures 9.1 and 9.2 plot 1/COP against 1/(cooling rate) for two chillers for: (*a*) the actual design value of ξ; and (*b*) the value of ξ at which the model predicts a globally-maximum COP (ξ = the fraction of the total heat rejection effected at the condenser). For *diagnostics*, consider Figure 9.2, where the measured COP of the chiller is significantly lower than the predicted maximum. The marked difference between theoretically-optimal and actual operating conditions stems from this single-stage chiller being severely limited by heat exchange

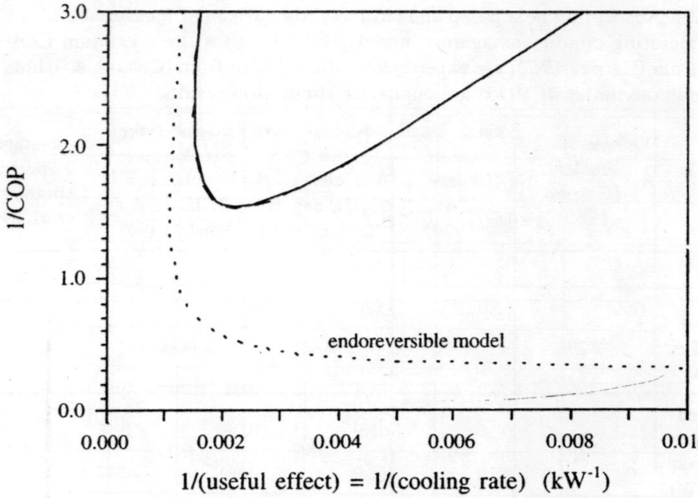

Figure 9.1: 1/COP plotted against 1/(cooling rate) for the single-stage absorption chiller reported in [Carrier 1962] for: (*a*) the actual design value of ξ = 0.44 (solid curve); (b) the value of ξ = 0.38 at which the model predicts a globally-maximum COP (broken curve); and (c) the endoreversible chiller model where only external losses are accounted for (dotted curve). Curves (a) and (*b*) are nearly indistinguishable. The control variable ξ is the fraction of total heat rejection effected at the condenser. Device parameters are summarized in Table 9.1.

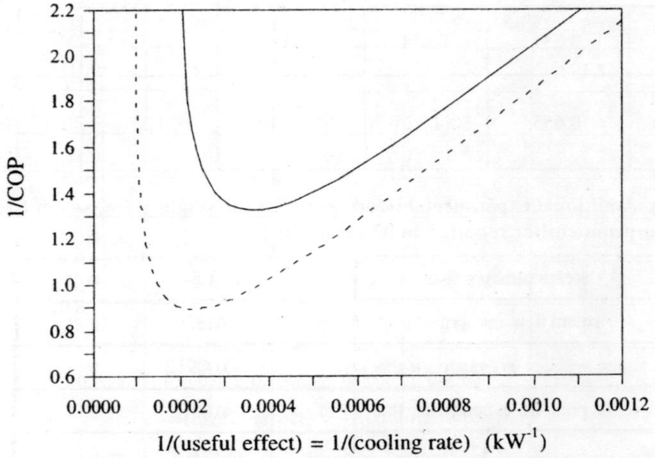

Figure 9.2: As in Figure 9.1, but computer simulation (rather than experimental) results for the far lower capacity single-stage absorption chiller reported in [Chuang & Ishida 1990]. Device parameters are listed in Table 9.1. Solid curve = simulated design with ξ = 0.473. Broken curve = model calculation with ξ = 0.08 at which a globally-maximum COP is predicted.

Table 9.2: Absorption heat pump and chillers: comparison of measured and simulated design operating conditions against model predictions for the maximum COP point. Figures from [Carrier 1962] are experimental data. Those from [Chuang & Ishida 1990] and [Abrahamsson *et al* 1995] are computer simulation results.

Variable ↓	single-stage chiller [Carrier 1962]	single-stage chiller [Chuang & Ishida 1990]	double-stage series cycle chiller [Chuang & Ishida 1990]	double-stage parallel-cycle chiller [Chuang & Ishida 1990]	single-stage heat pump [Abrahamsson *et al* 1995]
measured useful effect (kW)	529	4000	4000	4000	2410
calculated useful effect at maximum COP (kW)	449	4385	4556	4017	2417
measured ξ	0.380	0.473	0.291	0.327	0.447
calculated ξ at maximum COP	0.440	0.080	0.386	0.424	0.669
measured COP	0.634	0.664	1.17	1.32	1.71
calculated maximum COP	0.655	1.12	1.19	1.32	1.75

Table 9.3: Additional experimental information for the single-stage, steam-fired, LiBr–water absorption chiller reported in [Carrier 1962].

steam pressure (bar)	1.84
steam flow rate (kg s^{-1})	0.378
vapor pressure in evaporator (bar)	0.00912
vapor pressure in condenser (bar	0.0912
T_{evap} (°C)	5.6
T_{gen} (°C)	104.4
T_{abs} (°C)	41.7
T_{cond} (°C)	44.4

Table 9.4: Absorption heat transformers: summary of simulated performance data from [Abrahamsson *et al* 1995; Zhuo & Machielsen 1996].

Variable ↓	single-stage [Abrahamsson *et al* 1995]	single-stage [Zhuo & Machielsen 1996]	double-stage [Zhuo & Machielsen 1996]	triple stage [Zhuo & Machielsen 1996]	compact triple stage [Zhuo & Machielsen 1996]
T_{gen}^{in} (°C)	98	105	105	105	120
T_{evap}^{in} (°C)	105	105	105	105	120
T_{abs}^{in} (°C)	115	150	200	250	250
$(UA)_{gen}$ (kW K^{-1})	0.380	201	436	681	517
$(mCE)_{cond}$ (kW K^{-1})	0.111	150	313	483	645
$(UA)_{evap}$ (kW K^{-1})	0.600	398	442	487	1190
$(mCE)_{abs}$ (kW K^{-1})	0.238	200	200	200	200
ΔS_{int} (kW K^{-1})	0.000442	0.170	0.467	0.837	0.889

irreversibilities in the generator and condenser. Furthermore, the relatively high temperature stream is more suitable thermodynamically to double and triple-stage chillers (with their superior heat regeneration) than to single-stage chillers. At the calculated value of cooling rate where COP is globally maximized, the actual operating curve is dominated by heat exchange losses, and COP decreases rapidly as cooling rate decreases. Before configuring an absorption machine, one has the flexibility of varying: (*a*) the type of working fluid (*e.g.*, ammonia–water, LiBr–water, *etc.*); (*b*) the network connection scheme (*e.g.*, parallel or series; generator–absorber heat exchanger, dephlegmator, rectifier, *etc.*); and (*c*) internal regenerative heat transfer areas; among others. Once these elements are selected and sized, the distribution of irreversibilities will be a function of circulation flow rates and reservoir temperatures. During the operation of a given absorption unit, the irreversibilities influence one another. However at the design stage, prior to device construction, the assorted irreversibilities can be viewed as de-coupled in the sense that the simple thermodynamic model pertains to an arbitrary working fluid, network scheme and internal regeneration.

Table 9.5: Absorption heat transformers: comparison of simulated design operating conditions and model predictions for the maximum COP point.

variable ↓	single stage [Abrahamsson *et al* 1995]	single stage [Zhuo & Machielsen 1996]	double stage [Zhuo & Machielsen 1996]	triple stage [Zhuo & Machielsen 1996]	compact triple stage [Zhuo & Machielsen 1996]
measured useful effect (kW)	3.80	1000	1000	1000	1000
calculated useful effect at max. COP (kW)	2.36	987	1260	1447	1441
measured ψ	0.475	0.456	0.622	0.700	0.420
calculated ψ at max. COP	0.164	0.335	0.497	0.583	0.303
measured COP	0.475	0.455	0.285	0.206	0.162
calculated max. COP	0.643	0.457	0.293	0.218	0.174

B4. HEAT TRANSFORMER ANALYSIS AND DIAGNOSTICS

Consider the four heat transformers studied by [Zhuo & Machielsen 1996]. The single and double-stage heat transformers appear to have been designed relatively near their theoretically-optimal configurations. The deviation from nominally optimal operation increases as more stages are introduced (moving from single- to triple-stage units). In these particular installations, the single-stage heat transformer formed the modular building block for the double and triple-stage designs. Although the single-stage units appear to be properly optimized, the coupling between the building blocks is not. This is why the COP worsens with the number of stages (a trend unlike that in regenerative absorption machines that are designed *a priori* as double or multiple-stage machines). Table 9.6 summarizes the COP values, along with the relative contribution of external and internal losses, for the single, double and triple-stage heat transformers cited. The balance between external and internal losses is noticeably different for the single-stage heat

Table 9.6: Absorption heat transformers: relative contribution of external losses and internal losses to 1/COP (based upon simulation results from [Abrahamsson *et al* 1995; Zhuo & Machielsen 1996]). Heat leaks are negligible.

variable ↓	single stage [Abrahamsson *et al* 1995]	single stage [Zhuo & Machielsen 1996]	double stage [Zhuo & Machielsen 1996]	triple stage [Zhuo & Machielsen 1996]	compact triple stage [Zhuo & Machielsen 1996]
1/COP	2.11	2.20	3.51	4.86	6.16
contribution of external (heat transfer) losses	83.6%	81.7%	68.1%	58.9%	52.3%
contribution of internal losses	16.4%	18.3%	31.7%	41.1%	47.5%

*Due to round-off error in [Zhuo & Machielsen 1996], these reported energy flows do not sum exactly to 100%.

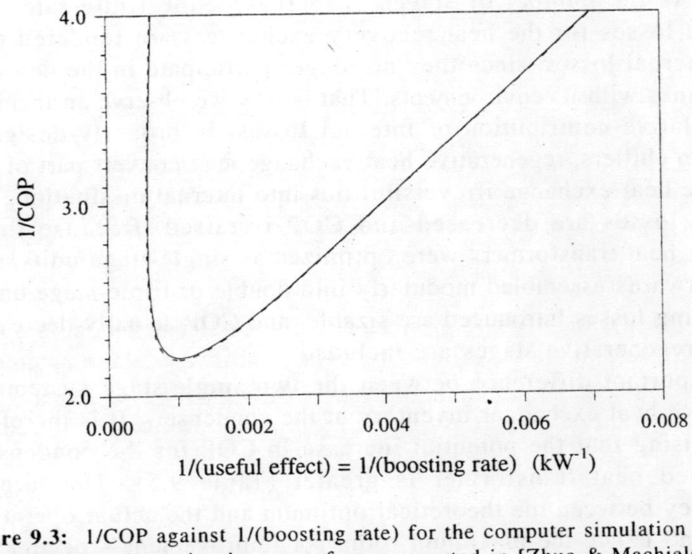

Figure 9.3: 1/COP against 1/(boosting rate) for the computer simulation results of the single-stage absorption heat transformer reported in [Zhuo & Machielsen 1996] for: (*a*) the actual design value of $\psi = 0.456$ (solid curve); and (*b*) the value of $\psi = 0.335$ at which the model predicts a globally-maximum COP (broken curve). The two curves are barely distinguishable. The control variable ψ is the fraction of total heat input effected at the generator. Device parameters are summarized in Table 9.4.

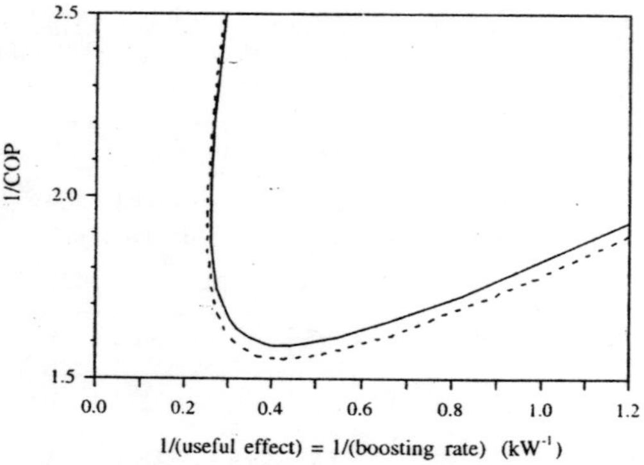

Figure 9.4: As in Figure 9.3, but for the computer simulation results of the single-stage absorption heat *transformer* reported in [Abrahamsson *et al* 1995]. Device parameters are summarized in Table 9.4. Solid curve = simulated design with $\psi = 0.475$. Broken curve = model calculation with $\psi = 0.164$ at which a globally-maximum COP is predicted.

transformers, relative to the other absorption and mechanical systems we've considered so far. Specifically, external losses dominate – in this instance because of the unusually large temperature differences across the heat exchangers (and not because of particularly small internal losses). As the number of stages is increased, the finite-rate heat exchange losses for the heat recovery exchangers are rendered part of the internal losses, since they no longer participate in the device's thermal links with its environments. That is why we observe an increase in the relative contribution of internal losses. In properly-designed absorption chillers, regenerative heat exchange may convert part of the finite-rate heat exchange irreversibilities into internal dissipation, but the total losses are decreased and COP is raised. Because these particular heat transformers were optimized as single-stage units, and only afterwards assembled modularly into double or triple-stage units, the coupling losses introduced are sizable, and COP actually decreases as more regenerative stages are included.

An important difference between the two single-stage systems is the relative heat exchanger inventory at the condenser. It is therefore not surprising that the potential increase in COP for the condenser-constrained heat transformer is greater (Table 9.5). The largest discrepancy between the theoretical optimum and the actual operating point – and hence the maximum room for improvement – occurs for the compact triple-stage heat transformer. The reduction in heat and mass transfer effectiveness necessitated by the compact design results both in enhanced internal dissipation and in a more severe heat exchanger thermal bottleneck.

Chapter 10

QUASI-EMPIRICAL THERMODYNAMIC MODEL FOR CHILLERS

"A little inaccuracy sometimes saves tons of explanation." - Saki (H.H. Munro)

A. INTRODUCTION

This chapter is devoted to a quasi-empirical thermodynamic chiller model that is easy to implement, and hence attractive for use in chiller predictive and diagnostic work. The original works are [Gordon & Ng 1994a; Gordon & Ng 1994b; Gordon & Ng 1995; Gordon *et al* 1995]. The quasi-empirical model has already been adopted in field studies and new standards in energy conservation in buildings [ASHRAE 1997] and in the refrigeration industry [Brandemuehl 1995].

As in Chapters 4 and 5, we adopt a blackbox perspective toward the chiller: it is a cooling machine which we are permitted to probe only through external non-intrusive measurements of variables such as power input, cooling rate, coolant flow rates and coolant temperatures. With just a handful of such measurements, we would like to fully characterize the chiller such that we can predict its thermodynamic performance for any realistic set of operating variables, and can diagnose potential problems signaled by a worsening of COP. We also reiterate that these models are valid only for *steady-state* chiller operation.

Like the fundamental model developed and explored in Chapters 4-9, the final result is a simple 3-parameter formula for the chiller performance curve. In fact, the regression procedure for the quasi-empirical model is even simpler and more robust than for the fundamental model. However, unlike the fundamental model, each of the parameters cannot be assigned a single clear physical significance. Hence the model parameters cannot be checked or determined through independent experimental measurement. The parameters include contributions from a mixture of physical mechanisms, and the relative contribution of each individual source of irreversibility cannot be ascertained. This is why we refer to the model as quasi-empirical: although the thermodynamic equations and approximations invoked in deriving the chiller performance curve are clear, the 3 parameters that characterize the chiller are empirical.

In the chronological development of accurate but general thermodynamic models for chillers, the quasi-empirical approach was proposed prior to the development of the fundamental models of Chapters 4-9. Given the predictive and diagnostic power of the quasi-empirical model for reciprocating, centrifugal and absorption chillers, as well as its calculational ease, we felt it worthwhile to devote a chapter to explaining and deriving it, and to illustrating its implementation for a wide range of real commercial chillers.

For absorption chillers in particular, the quasi-empirical model offers the advantage over the fundamental model of permitting comparisons between measured and predicted chiller performance with the limited data usually reported in manufacturer catalogs. Manufacturers usually do not report all 4 coolant temperatures (generator, condenser, evaporator and absorber); neither do they report the division of total heat rejection between the condenser and absorber. The approximations invoked in the quasi-empirical model provide a predictive tool based solely on data for: cooling rate, COP, T^{in}_{gen}, and T^{in}_{cond} and T^{in}_{evap}.

Recall that this is not the case for the fundamental absorption chiller model of Chapters 4, 5 and 9. In those instances, the absorption chiller model was used for determining optimal configurations and comparing them to commercial units, but based on data for a single operating point. Here we can test model predictions directly over a broad range of cooling rates. However the quasi-empirical model does not offer the ability to optimize the chiller configuration.

We will derive the quasi-empirical model, explain how its parameters are determined from chiller performance data, and then demonstrate the power of the model with actual performance data. For reciprocating chillers, those measurements are from manufacturer catalogs and represent an extensive range of chiller size, COP values and coolant temperatures. For centrifugal chillers, we will offer a case study where deterioration in the performance of a large commercial installation was diagnosed without any intrusive measurements. For absorption chillers, laboratory measurements on a commercial unit and manufacturer catalog data will be used to test the trends and the accuracy of model predictions. A case study on how the introduction of surfactant ameliorates absorption chiller performance will also be presented.

B. DERIVATION OF THE MODEL FOR MECHANICAL CHILLERS

B1. Energy and entropy balance

We start with the analysis of mechanical chillers. The First Law for the change in the internal energy of the refrigerant in the chiller cycle is expressed as

$$\Delta E = 0 = P_{in} - Q_{cond} - Q_{evap}. \tag{10.1}$$

The quasi-empirical model differs from the more fundamental approach in the modeling of internal dissipation (and heat leaks). Rather than lumping all internal losses into one constant entropy production term as in Equation (4.16), the internal losses at the hot (condenser) and cold (evaporator) ends are treated as modifying the respective heat transfer rates. The Second Law for the change in the entropy of the refrigerant can then be expressed as

$$\Delta S = 0 = \frac{Q_{cond} + q_{cond}}{T_{cond}} - \frac{Q_{evap} + q_{evap}}{T_{evap}} \tag{10.2}$$

where q_{cond} and q_{evap} are the additional heat transfers that stem from internal losses. Combining Equations (10.1) and (10.2) with the definition of COP, we obtain

$$\frac{1}{\text{COP}} = -1 + \frac{T_{cond}}{T_{evap}} + \frac{\left[\dfrac{q_{evap} T_{cond}}{T_{evap}} - q_{cond}\right]}{Q_{evap}}. \tag{10.3}$$

B2. Heat exchanger effects: expressing results in terms of coolant temperatures

Since chiller performance is invariably reported as a function of the readily measurable coolant (as opposed to refrigerant) temperatures, we need to introduce the heat exchanger equations in order to arrive at a final easy-to-implement performance formula. With manufacturer catalog data for reciprocating and centrifugal chillers usually reported in terms of T_{cond}^{in} and T_{evap}^{out}, we express the heat exchanger energy balance equations as

$$T_{\text{cond}} = T_{\text{cond}}^{\text{in}} + \frac{Q_{\text{cond}}}{(mCE)_{\text{cond}}} \equiv T_{\text{cond}}^{\text{in}} + \frac{Q_{\text{cond}}}{M_{\text{cond}}} \tag{10.4}$$

$$T_{\text{evap}} = T_{\text{evap}}^{\text{out}} - \frac{Q_{\text{evap}}(1 - E_{\text{evap}})}{(mCE)_{\text{evap}}} \equiv T_{\text{evap}}^{\text{out}} - \frac{Q_{\text{evap}}}{M_{\text{evap}}} \tag{10.5}$$

where $M_{\text{cond}} = (mCE)_{\text{cond}}$ and $M_{\text{evap}} = (mCE)_{\text{evap}}$ are shorthand notation for the thermal throughput at the condenser and evaporator, respectively.

In combining Equations (10.3)–(10.5) and solving for COP in terms of cooling rate and chiller parameters, it is convenient to express the result as an expansion about the limit of large M_{cond} and M_{evap}. Properly operating commercial chillers operate over a range of cooling rates for which internal irreversibilities are dominant. Solving for 1/COP, we obtain the leading term $1/\text{COP}_o$ (zeroth order in $1/M_{\text{cond}}$ and $1/M_{\text{evap}}$) as

$$\frac{1}{\text{COP}_o} = -1 + \left[\frac{T_{\text{cond}}^{\text{in}}}{T_{\text{cond}}^{\text{out}}}\right] + \frac{\left[\dfrac{q_{\text{evap}} T_{\text{cond}}^{\text{in}}}{T_{\text{evap}}^{\text{out}}} - q_{\text{cond}}\right]}{Q_{\text{evap}}}. \tag{10.6}$$

The contributions to the term of first order in $1/M$, denoted by $1/\text{COP}_1$, can be grouped according to cooling-rate dependence:

$$\begin{aligned} \frac{1}{\text{COP}_1} = & \left[\frac{1}{Q_{\text{evap}}}\right]\left[\frac{q_{\text{evap}}}{M_{\text{cond}} T_{\text{evap}}^{\text{out}}}\right]\left[\frac{q_{\text{evap}} T_{\text{cond}}^{\text{in}}}{T_{\text{evap}}^{\text{out}}} - q_{\text{cond}}\right] + \\ & \left[\frac{Q_{\text{evap}}}{T_{\text{evap}}^{\text{out}}}\right]\left[\frac{T_{\text{cond}}^{\text{in}}}{T_{\text{evap}}^{\text{out}}}\right]\left[\frac{1}{M_{\text{cond}}} + \frac{1}{M_{\text{evap}}}\right] + \\ & \frac{\dfrac{q_{\text{cond}}}{M_{\text{evap}}} + \dfrac{q_{\text{evap}} T_{\text{cond}}^{\text{in}}}{T_{\text{evap}}^{\text{out}} M_{\text{cond}}} + \left[\dfrac{T_{\text{cond}}^{\text{in}} q_{\text{evap}}}{T_{\text{evap}}^{\text{out}}} - q_{\text{cond}}\right]\left[\dfrac{1}{M_{\text{cond}}} + \dfrac{1}{M_{\text{evap}}}\right]}{T_{\text{evap}}^{\text{out}}}. \end{aligned} \tag{10.7}$$

The first term on the right-hand side of Equation (10.7), *i.e.*, the $1/Q_{\text{evap}}$ term, turns out to be a negligible correction to the $1/Q_{\text{evap}}$ term in Equation (10.6) (at least based upon performance data from commercial reciprocating and centrifugal chillers). The second term on the right-hand side of Equation (10.7), of order Q_{evap}, can in principle play an important role in chillers where heat-exchanger irreversibilities are significant; but this turns out not to be the case for the wide range of commercial mechanical chillers we examined. The third term (independent of cooling rate), which we denote by h_X,

$$h_X = \frac{\dfrac{q_{cond}}{M_{evap}} + \dfrac{q_{evap} T_{cond}^{in}}{T_{evap}^{out} M_{cond}} + \left[\dfrac{T_{cond}^{in} q_{evap}}{T_{evap}^{out}} - q_{cond}\right]\left[\dfrac{1}{M_{cond}} + \dfrac{1}{M_{evap}}\right]}{T_{evap}^{out}} \tag{10.8}$$

is dimensionless and typically turns out to be small (compared to unity). However, when noticeable heat exchanger fouling occurs, for example, the h_X term can grow to be non-negligible. Also, in cases where the only changes in chiller behavior stem from changes in the heat exchangers, and where a diagnostic capability is desired, such that chiller performance is being compared at similar coolant temperatures but different heat exchanger conditions, it becomes important to retain the h_X term.

Hence to a good approximation, one can express chiller COP as

$$\frac{1}{\text{COP}} = -1 + \left[\frac{T_{cond}^{in}}{T_{evap}^{out}}\right] + \frac{\left[\dfrac{q_{evap} T_{cond}^{in}}{T_{evap}^{out}} - q_{cond}\right]}{Q_{evap}} + h_X . \tag{10.9}$$

Terms that are higher order in Q_{evap} turn out to be negligible.

There are situations in which the cooling-rate dependence due to finite-rate heat transfer grows noticeable. This contingency will be addressed for reciprocating chillers in Section C. The onset of nonlinearities in the chiller characteristic curve can also be observed in manufacturer catalog data for *absorption* chillers (presented in detail in Section E). They also form a considerable part of the measurable performance curve for the thermoacoustic and thermoelectric refrigerators, as shown in Section F. For now, we proceed subject to the assumption that 1/COP is well approximated as a linear function of $1/Q_{evap}$.

B3. Modeling internal losses and the final 3-parameter formula

The approximate derived relation is then

$$\frac{1}{\text{COP}} = -1 + \left[\frac{T_{\text{cond}}^{\text{in}}}{T_{\text{evap}}^{\text{out}}}\right] + \frac{\left[\dfrac{q_{\text{evap}} T_{\text{cond}}^{\text{in}}}{T_{\text{evap}}^{\text{out}}} - q_{\text{cond}}\right]}{q_{\text{evap}}}. \tag{10.10}$$

An accurate model must account for the particular functional dependences of the loss terms q_{cond} and q_{evap} on coolant temperatures in order to yield a final utilizable expression for COP as a function of Q_{evap} and coolant temperatures only. We invoke the approximation that these loss terms are thermodynamically linear, in that they can be expanded about reversible behavior up to first order in temperature differences based on the corresponding refrigerant temperature:

$$q_{\text{cond}} = -A_o + A_3 T_{\text{cond}} \tag{10.11}$$

$$q_{\text{evap}} = -A_2 + A_4 T_{\text{evap}} \tag{10.12}$$

where the A's are constants. Using Equations (10.11) and (10.12) in (10.10), and defining $A_1 = A_3 + A_4$, one obtains

$$\frac{1}{\text{COP}} = -1 + \frac{T_{\text{cond}}^{\text{in}}}{T_{\text{evap}}^{\text{out}}} + \frac{A_1 T_{\text{cond}}^{\text{in}} - A_o - A_2\left(T_{\text{cond}}^{\text{in}} / T_{\text{evap}}^{\text{out}}\right)}{Q_{\text{evap}}} \tag{10.13}$$

where the constants A_o, A_1 and A_2 characterize the internal irreversibilities of a particular chiller.

The 3 parameters A_o, A_1 and A_2 relate to dissipative losses in a manner that does not permit their independent experimental measurement. This stands in contrast to the fundamental chiller model of Chapters 4 and 5 where the identity of each of the chiller parameters was definitively linked to a particular physical mechanism and could hence be checked via independent experimental measurement. This somewhat unsatisfying aspect of the quasi-empirical model does not detract from its ability to accurately capture the behavior of real commercial mechanical chillers.

Note that in cases where the coolant temperatures are roughly constant, while cooling rate varies (via changes in the refrigerant temperature), the characteristic plot of 1/COP against $1/Q_{\text{evap}}$ should be a single straight line. In other words, the cooling rate dependence of the COP in the operating regime of interest should stem primarily from internal losses.

In Section C, we will first illustrate the *predictive* power of the quasi-empirical model with extensive performance data for reciprocating chillers from manufacturer catalogs. To highlight the *diagnostic* value of the model, we present a case study for a centrifugal chiller in Section D. Section E extends the quasi-empirical model to absorption machines. Finally, Section F examines the quasi-empirical model as applied to less conventional chillers.

C. RECIPROCATING CHILLERS

C1. Validating predicted functional dependences and accurate COP correlations

As we saw in Chapter 5, manufacturers usually provide performance data for reciprocating chillers in the form of a plethora of measurements that cover a broad range of coolant temperatures. That enables the user to predict chiller performance over almost any reasonable anticipated operating conditions. This experimental procedure is quite time-consuming. Completely empirical fits to the results may not be extrapolated beyond the range of conditions for which measurements were carried out. It would be preferable to be able to fully characterize a given chiller from just a handful of measurements.

The quasi-empirical model provides such an opportunity. A small number of judiciously selected measurements permit one to regress for the 3 parameters A_o, A_1 and A_2 that characterize the chiller, and then to use Equation (10.13) to predict chiller COP for any anticipated set of coolant temperatures and cooling rate.

As an illustration, we show in Figures 10.2–10.4 the model's accuracy in correlating the basic functional dependences of key system variables for one representative commercial reciprocating chiller. The data are listed in Table 10.1. The model predicts that 1/COP should be a linear function of $1/Q_{\text{evap}}$ and two additional independent variables, which can be taken as $T_{\text{cond}}^{\text{in}}$ and the ratio $T_{\text{cond}}^{\text{in}} / T_{\text{evap}}^{\text{out}}$.

Note that a plot of

$$\left\{\frac{1}{\text{COP}} + 1 - \frac{T_{\text{cond}}^{\text{in}}}{T_{\text{evap}}^{\text{out}}}\right\} Q_{\text{evap}} \text{ against } \frac{T_{\text{cond}}^{\text{in}}}{T_{\text{evap}}^{\text{out}}} \qquad (10.14)$$

should yield a set of parallel straight lines of slope $-A_2$, each line for a different value of $T_{\text{cond}}^{\text{in}}$ (see Figure 10.1). To check this prediction thoroughly, we analyzed data not only from the chiller noted in Table

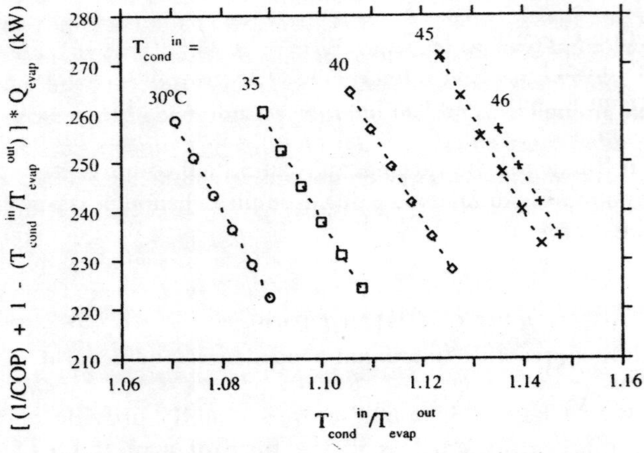

Fig.10.1 Partial check of the predicted functional dependence of Equation (10.13), cast in the form of (10.14), for the chiller summarized in Table 10.1. The model predicts a series of straight lines of the same slope, independent of T^{in}_{cond}. The 5 lines have slopes that vary by $\pm$2.2%. The constant slope value is the model parameter $-A_2$.

10.1, but also 29 other commercial reciprocating chillers summarized in Table 10.2. These 30 chillers are mass-produced efficient chillers used in large central space cooling applications, and use the refrigerant Freon R-12. We found that each case reconfirms the model prediction illustrated in Figure 10.1.

Having ascertained the value of A_2, we note that in a plot of

$$\left\{\frac{1}{\text{COP}}+1-\frac{T^{\text{in}}_{\text{cond}}}{T^{\text{out}}_{\text{evap}}}\right\}Q_{\text{evap}}+\frac{A_2 T^{\text{in}}_{\text{cond}}}{T^{\text{out}}_{\text{evap}}} \quad \text{against} \quad T^{\text{in}}_{\text{cond}} \tag{10.15}$$

all data points should fall on a single straight line with slope A_1 and ordinate-intercept $-A_o$. Experimental confirmation of this prediction is illustrated in Figure 10.2 for the chiller of Table 10.1, and was also successfully checked for the other 29 chillers listed in Table 10.2.

When the predictive ability of the quasi-empirical model was tested with data from the 30 commercial reciprocating chillers cited above, it was found to correlate measured COP values, over the full range of operating conditions covered in the manufacturers' catalogs (T^{out}_{evap} = 4.4–15.0°C and T^{in}_{cond} = 23.9–53.0°C), with a rms error of 0.4%, for around 900 data points (see Figure 10.3 and Table 10.2). The rms error is well below the experimental uncertainty of ±3% in the measurements reported.

Table 10.1: Manufacturer catalog data (28 measured points) for the reciprocating chiller analyzed in Figures 10.2 and 10.3 [Trane 1992]. The nominal rated cooling rate is 1172 kW.

T_{cond}^{in} (K)	T_{evap}^{out} (K)	Q_{evap} (kW)	COP
303	278	1075	3.37
303	279	1107	3.41
303	280	1139	3.45
303	281	1172	3.50
303	282	1205	3.54
303	283	1239	3.58
308	278	1014	3.04
308	279	1045	3.08
308	280	1076	3.11
308	281	1108	3.15
308	282	1140	3.19
308	283	1172	3.22
313	278	954	2.74
313	279	982	2.77
313	280	1013	2.81
313	281	1042	2.83
313	282	1073	2.86
313	283	1103	2.89
318	278	892	2.47
318	279	920	2.49
318	280	948	2.52
318	281	976	2.54
318	282	1004	2.56
318	283	1032	2.58

Table 10.1: continued

319	278	880	2.41
319	279	907	2.44
319	280	934	2.46
319	281	962	2.49

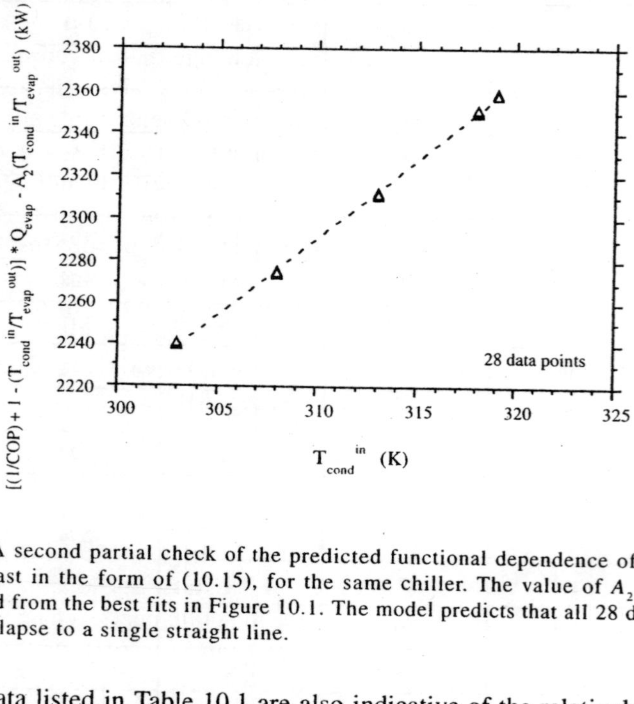

Fig.10.2 A second partial check of the predicted functional dependence of Equation (10.13), cast in the form of (10.15), for the same chiller. The value of A_2 used was determined from the best fits in Figure 10.1. The model predicts that all 28 data points should collapse to a single straight line.

The data listed in Table 10.1 are also indicative of the relatively narrow range of cooling rates of which reciprocating chillers are capable without unloading piston–cylinder units: down to around 70% of the maximum (75% of the rated cooling rate).

C2. Limits to the model

Understanding the physics and approximations underlying the quasi-empirical model permits us not only to predict when the model should provide accurate predictions, but also when the model should cease to do so. The onset of noticeable deviations from the linearity of the curve of 1/COP against 1/(cooling rate) should occur when finite-rate heat transfer at the heat exchangers becomes a significant bottleneck. The temperature differences across each chiller component are then larger, the COPs are lower, and the higher-order corrections in orders of Q_{evap} in Equation (10.7) grow in magnitude. We should see this in a gradual worsening of the functional dependences predicted by Equation (10.13) and poorer accuracy in predicting chiller COP. In terms of the characteristic performance curve, the chiller is moving off the linear section toward *relatively* high cooling rate.

To test these claims, we analyzed manufacturer catalog data for a nominal 63 kW cooling rate *air*-cooled reciprocating chiller, with COPs in the range 1.5–3.5, and cooling rates of 37–65 kW [Carrier 1984]. Air cooling usually means poorer heat exchange and therefore a greater contribution of external losses. The results for analyses corresponding to those presented earlier in this section are presented in Figures 10.5–10.7. The functional dependence expected from (10.14) is tested in Figure 10.4. Unlike Figure 10.1, there are small but noticeable differences in the slope of each of the straight lines. Furthermore, as shown in Figure 10.5, the predicted functional dependence of (10.15) worsens. All data points do not collapse into a single linear relation (contrast with Figure 10.2). Finally, obtaining the best 3-parameter fit of Equation (10.13) and summing over all 24 data points, we find that the predicted COPs have a rms error of 4% relative to measured values (see Figure 10.6, and compare with Figure 10.3).

D. CENTRIFUGAL CHILLERS

D1. Details of a diagnostic case study

Whereas our analysis for reciprocating chillers emphasized the predictive power of the quasi-empirical model, we will use a case study for a large commercial centrifugal chiller to illustrate the diagnostic value of the model. In this particular diagnostic study, heat exchanger fouling plays a key role. Hence, on the right-hand side of Equation (10.9), we retain the h_X term. The key point to be stressed is that a plot of 1/COP against $1/Q_{evap}$ should be linear, and that only the (extrapolated) ordinate intercept, and not the slope, should depend on heat exchanger properties.

The air-conditioning plant we monitored is described in detail in [Gordon *et al* 1995]. The chiller's nominal rated cooling rate was 352 kW [Carrier 1971]. *In situ*, steady-state measurements were made on the installed chiller for a period of 6 months. After 6 months, in view of the chiller's COP having decreased to undesirably low levels, maintenance was performed and the heat exchanger tubes were cleaned. Chiller COP then increased by 36% on average, at roughly the same values of coolant temperatures and at approximately the same values of cooling rate (see Table 10.3).

Table 10.2 Data for the 30 reciprocating chillers analyzed. The ranges of variables covered are: (a) rated cooling rate = 30–1300 kW; (b) T_{evap}^{out} = 4.4–15.0°C; (c) T_{cond}^{in} = 23.9–53.0°C; and (d) COP = 2.33–6.29.

Chiller model	Reference	Rated capacity (kW)*	Number of data points	rms error in correlating COP (%)
Carrier 30HKA015	Toyo 1991	47.60	25	0.49
Carrier 30HKA020	Toyo 1991	64.00	25	0.43
Carrier 30HKA030	Toyo 1991	82.40	25	0.61
Carrier 30HK040	Toyo 1991	119.0	25	0.48
Carrier 30HK050	Toyo 1991	160.0	25	0.44
Carrier 30HK060	Toyo 1991	192.0	25	0.43
Carrier 30HK080	Toyo 1991	238.0	25	0.50
Carrier 30HK100	Toyo 1991	315.0	25	0.44
Carrier 30HK120	Toyo 1991	358.0	25	0.51
Carrier 30HR140	Toyo 1991	440.0	25	0.54
Carrier 30HR160	Toyo 1991	481.0	25	0.60
Trane CGAV214	Trane 1992	585.8	28	0.33
Trane CGAV422	Trane 1992	852.4	30	0.33
Trane CGAV426 o/s	Trane 1992	930.6	38	0.46
Trane CGAV426	Trane 1992	1065	30	0.28
Trane CGAV428 #	Trane 1992	1172	28	0.32
Trane CGAV430	Trane 1992	1279	28	0.41
Trane CCAC 70R C60	Trane 1990	242.2	35	0.29
Trane CCAC 70R C80	Trane 1990	255.6	35	0.29
Trane CCAC 80R C80	Trane 1990	255.6	35	0.30
Trane CGWC C70R	Trane 1990	257.7	35	0.26
Trane CCAC 80R D10	Trane 1990	295.33	35	0.31
Trane CGWC C80R	Trane 1990	296.4	35	0.20
Trane CGWC C90R	Trane 1990	328.4	35	0.27
Trane CGWC D10R	Trane 1990	367.8	35	0.25
Trane CGWC D11R	Trane 1990	402.7	35	0.24
Trane CGWC D12R	Trane 1990	445.6	35	0.18
10 ton	Leverenz & Bergan 1983	36.20	30	0.66
20 ton	Leverenz & Bergan 1983	74.50	30	0.61
50 ton	Leverenz & Bergan 1983	178.6	30	0.31

*Rated cooling rate defined at T_{evap}^{out} = 10°C and T_{cond}^{in} = 35°C
#This is the chiller cited in Table 10.1 and Figures 10.2 and 10.3

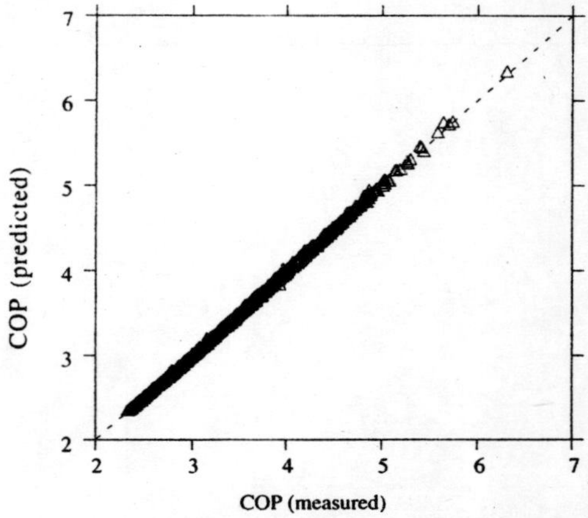

Fig.10.3: Illustration of the accuracy of the quasi-empirical model in correlating reciprocating chiller COP. Predicted COP is plotted against measured COP for the 30 chillers listed in Table 10.2, spanning nominal cooling rates from 30 to 1300 kW. A total of 897 data points are represented.

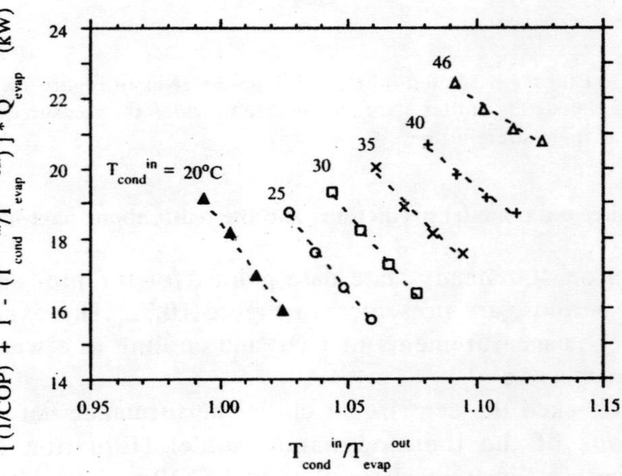

Fig.10.4: Same test for the predicted functional dependence as in Figure 10.1, but for the *air*-cooled reciprocating chiller. Note that the slopes of the lines change non-negligibly with T_{cond}^{in}.

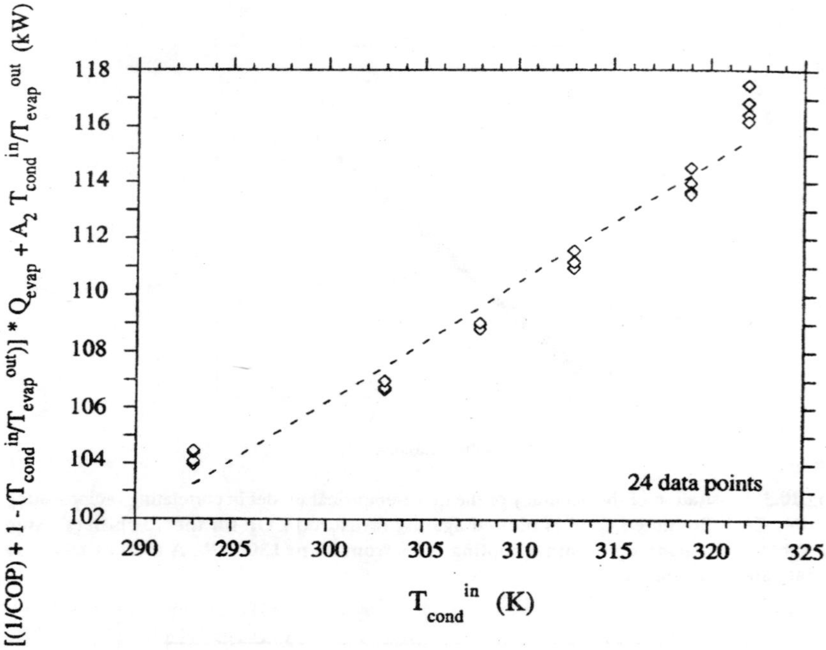

Fig.10.5 Same test for the predicted functional dependence as in Figure 10.2, but for the air-cooled reciprocating chiller. Note the degree to which the measured points do not collapse to a single straight line.

D2. Performance data, model predictions and the truth about part-load behavior

Results based on 400 steady-state data points, for the pre- and post-maintenance periods, are presented in Figure 10.7. The experimental uncertainty in measurements of COP and cooling rate was ±6.0% and ±4.5%, respectively.

First, we checked the centrifugal chiller performance data against the predictions of the thermodynamic model (Equation (10.9)). Figure 10.7 presents characteristic plots of 1/COP against $1/Q_{evap}$ that clearly illustrate the division into the pre- and post-maintenance periods. The distinction between full- and part-load periods is also evident.

Centrifugal chillers invariably exhibit COPs that increase with cooling rate. A common misinterpretation of centrifugal chiller rating conditions led to the engineering rule of thumb that COP is largest at part-load (typically 50–80% load) conditions [Kreider & Rabl 1994]. As demonstrated conclusively in [Austin 1991; Beyenem *et al* 1994; Liu *et al* 1994] with experimental data for a broad range of commercial

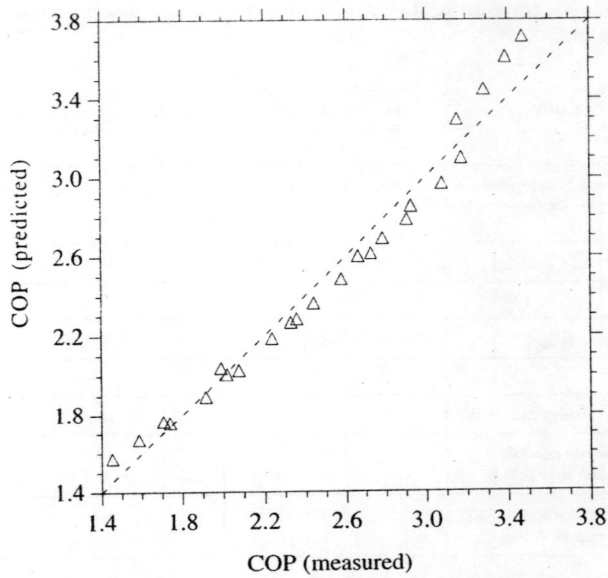

Fig.10.6 Illustration of the poorer accuracy of the quasi-empirical model in predicting COP for the air-cooled reciprocating chiller. Predicted COP is plotted against measured COP.

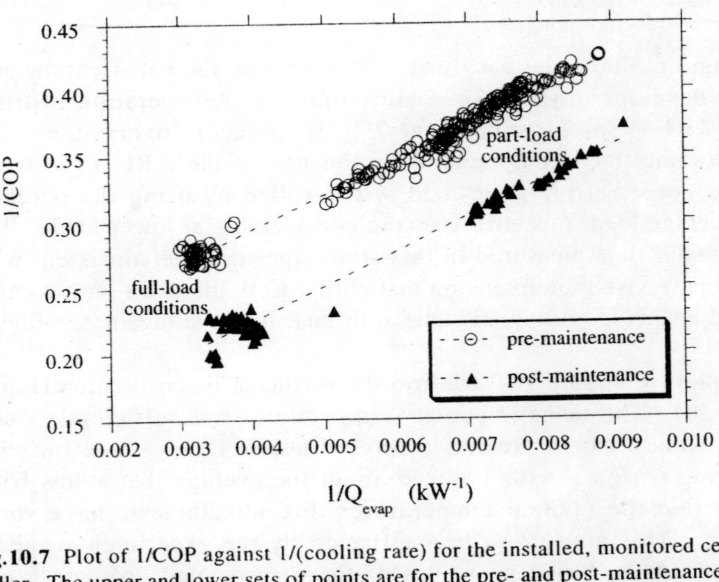

Fig.10.7 Plot of 1/COP against 1/(cooling rate) for the installed, monitored centrifugal chiller. The upper and lower sets of points are for the pre- and post-maintenance periods, respectively. Maintenance involved cleaning the condenser and evaporator heat exchanger tubes. Linear regression best fits are drawn.

Table 10.3 Summary of centrifugal chiller data for the pre- and post-maintenance periods [Gordon *et al*1995]

variable	pre–maintenance	post–maintenance
number of data points	278	122
average T_{evap}^{out} (°C)	6.7±0.8	6.6±1.5
average T_{cond}^{in} (°C)	29.5±0.8	28.8±0.8
average cooling rate (kW)	179	207
maximum recorded cooling rate (kW)	334	297
minimum recorded cooling rate (kW)	113	109
average COP	2.89	3.94
maximum recorded COP	3.76	5.13
minimum recorded COP	2.35	2.68

centrifugal chillers, the part-load COP based on the certification procedures developed by the Air-conditioning and Refrigeration Institute (ARI) [ARI 1988] for a constant T_{cond}^{in} is markedly overstated. The error is sometimes compounded by the use of the ARI curve when the percentage refrigeration load is controlled by using the percentage amperage load, this error growing considerably at lower loads. The experimental data measured in this study appear to be consistent with the correct, revised interpretation that chiller COP increases with cooling rate, and attains its maximum value at or near full load (again, see Figure 10.7).

The plots of Figure 10.7 confirm the predicted linear relation (Equation (10.9)). The range of coolant temperatures was sufficiently small that the simple model predicts a single linear relation – one for each monitoring period – with a spread about the average that stems from the fact that the coolant temperatures did, nonetheless, have some variation. This appears to be confirmed by the experimental measurements. Also, based on values of the extrapolated ordinate intercept, the h_x term in Equation (10.9) (and hence 1/COP) decreases by 0.07 from the pre- to the post-maintenance period.

The maintenance performed should not have affected the internal irreversibilities of the chiller. Rather, cleaning heat exchanger tubes should only affect the irreversibilities associated with finite-rate heat transfer (the h_x term in Equation (10.9)). The refrigerant temperatures changed (condenser temperature decreasing and evaporator temperature increasing), while the chiller operated at the same coolant temperatures. In terms of model predictions, the pre- and post-maintenance periods should exhibit 1/COP versus $1/Q_{evap}$ plots with the same slope but different extrapolated ordinate intercepts. This is precisely what was observed.

D3. The diagnostic case study from the perspective of the fundamental chiller model

Although the diagnostic case study recounted above was originally carried out with the quasi-empirical model, the results can equally well be understood from the perspective of the fundamental chiller model derived in Chapters 4–6. The characteristic chiller performance formula, Equation (6.1), can be expanded in powers of Q_{evap}. The dominant terms are of order Q_{evap}^{-1} and Q_{evap}^{0}. Terms of order Q_{evap}^{1} and higher constitute relatively small contributions. The expression for 1/COP, in terms of the measured coolant temperatures T_{cond}^{in} and T_{evap}^{out} and the measured cooling rate Q_{evap} is

$$\frac{1}{\mathrm{COP}} = \left[-1+\frac{T_{cond}^{in}}{T_{evap}^{out}}\right]+\left[\frac{T_{cond}^{in}\Delta S_{int}}{Q_{evap}}\right]+\left[R\,\mathrm{Order}(Q_{evap}^{0})\right] + \left[\mathrm{Order}(Q_{evap}^{1})\right]+\ldots \tag{10.16}$$

(where heat leaks are taken to be negligible relative to internal dissipation, but could easily be included in the $1/Q_{evap}$ term if required).

The monitored centrifugal chiller experienced relatively constant coolant temperatures over a broad range of cooling rates. Hence, from Equation (10.16), a plot of 1/COP against $1/Q_{evap}$ should yield a straight line, the slope of which is proportional to the rate of internal dissipation, and the ordinate intercept of which is linear in the overall heat exchanger thermal resistance. The "before and after" data plotted in Figure 10.7, whereby chiller performance degradation produced an increase in ordinate intercept but no change in slope, then accede to the same interpretation at which we originally arrived from the viewpoint of the quasi-empirical model.

Tutorial 10.1

Commercial manufacturers of centrifugal chillers tend to present chiller performance at rated capacity, which does not reflect the true field or site performance. The Integrated-Part-Load Value (IPLV) approach proposed by ARI Standard 550 attempts to bridge this gap between the performance provided by manufacturers and the manner in which chillers are presumed to operate in the field. ARI Standard 550 stipulates that chiller performance, in the form of 1/COP (usually cited by chiller engineers in terms of kW per Rton), be divided into 4 bins of 0–25%, 25–50%, 50–75% and 75–100% of the chiller's installed capacity, and that each bin be assigned an equal weighting, *i.e.*, an equal frequency of occurrence.

A more accurate procedure is to conduct a survey of the actual frequency distribution (even for the coarse binning prescribed by the ARI Standard), weight each of the 4 periods correctly and calculate the correct average value of 1/COP.

In this tutorial, we examine the difference in long-term average 1/COP between the ARI Standard method and the correct procedure. We use the performance data presented in the diagnostic case study presented above in Sections D1 and D2. To an excellent approximation, the graphs of 1/COP against $1/Q_{evap}$ can be fit by the following linear relations for the pre- and post-maintenance periods:

$$1/\text{COP} = \frac{26.3}{Q_{evap}} + 0.189 \qquad \text{pre-maintenance}$$

$$1/\text{COP} = \frac{26.2}{Q_{evap}} + 0.120 \qquad \text{post-maintenance}$$

where Q_{evap} is in kW. In addition, the actual measured load frequency distribution was

frequency of occurence	range as a percentage of full load
0.00	0–25%
0.10	25–50%
0.25	50–75%
0.65	75–100%

and the maximum measured steady-state cooling load was 334 kW.

The calculations for 1/COP during each nominal period are based on the average value for Q_{evap} in each bin. The results are:

load range as percentage of maximum	ARI Standard 550 IPLV method		Based on measured load pattern	
	pre–maintenance	post–maintenance	pre–maintenance	post–maintenance
0–25%	0.819	0.748	0.819	0.748
25–50%	0.399	0.329	0.399	0.329
50–75%	0.315	0.246	0.315	0.246
75–100%	0.279	0.210	0.279	0.210
average 1/COP	0.453	0.383	0.300	0.231

Let's translate these figures into electricity costs for operating the chiller. First, consider the savings associated with the maintenance on the chiller; and second, consider the annual cost of running the chiller. 1/COP is the convenient variable for estimating operating costs since, when multiplied by the average cooling load, it yields the average electricity consumption for the chiller.

We take a period of one year with 24 hour a day, 7 day a week operation (8760 operating hours per year). Our estimate of yearly-average electricity costs is based on the yearly-average cooling load. The frequency-weighted yearly-average cooling load is 255 kW. The IPLV yearly-average cooling load is 167 kW. Take an electricity cost of \$0.10 kWh^{-1}.

The predicted savings linked to the chiller maintenance are estimated by the IPLV method to be

$$(\$0.10)\ (8760)\ (0.453 - 0.383)\ (167) = \$1.02 \times 10^4$$

The actual savings are estimated as

$$(\$0.10)\ (8760)\ (0.300 - 0.231)\ (255) = \$1.54 \times 10^4$$

For the post-maintenance chiller, the annual operating costs from the IPLV method are

$$(\$0.10)\ (8760)\ (0.383)\ (167) = \$5.60 \times 10^4$$

while the actual operating costs are estimated as

$$(\$0.10)\ (8760)\ (0.231)\ (255) = \$5.16 \times 10^4.$$

E. ABSORPTION CHILLERS

E1. Basic thermodynamic behavior

Next, we attempt to extend the quasi-empirical model to absorption chillers. For the basics of the operation and modeling of absorption chillers, refer back to Chapters 2, 4 and 9. Detailed thermodynamic models for absorption chillers are highly device-specific. They require a large number of input material and component parameters. We'll see shortly that, at least for single-stage absorption chillers, the quasi-empirical model reduces to a simple analytic formula, and compares favorably with experimental performance data.

In double (or multiple) stage absorption chillers, *all* the losses in the heat recovery exchangers are *internal* losses, because they are not part of the chiller's thermal communication with its surroundings. This means that the irreversibility of finite-rate heat exchange in the heat recovery exchangers is *internal* dissipation. In mechanical chillers, and in single-stage absorption chillers, finite-rate heat exchange constitutes *external* losses. This point goes beyond a simple semantic distinction. The functional dependence of finite-rate heat exchange irreversibilities on cooling rate is markedly different from that of mechanisms such as fluid friction and mass-transfer resistive losses.

Hence, whereas single-stage absorption chillers may have internal dissipation rates that are well approximated as constant, double (and triple) stage absorption chillers exhibit significant deviations from such behavior. (A quantitative example is presented in Chapter 14.) The extension of the quasi-empirical model to absorption chillers relies upon retaining the approximation of a roughly constant rate of internal dissipation. Therefore its validity should be restricted to single-stage machines.

In terms of the characteristic chiller performance curve of 1/COP against $1/Q_{evap}$, single-stage absorption chillers, like mechanical chillers, exhibit a linear regime at relatively low cooling rates. But performance data for commercial absorption chillers also usually extend to cooling rates beyond the point of maximum COP, and can encounter a point where cooling rate peaks (recall Figure 5.1, and see Figure 10.11 below). The physics belying this behavior at relatively high cooling rates – in contrast to the absence of such behavior for mechanical chillers – was covered in Chapters 4 and 9, and stems from the chiller being driven by heat rather than work. The point will not be belabored. Rather, we will derive the quasi-empirical model for absorption chillers, and test its predictions against actual performance data.

E2. Adapting the quasi-empirical model to absorption chillers

Absorption chillers have the same type of irreversibilities in the condenser and evaporator as mechanical chillers. We need to embellish the model with losses at the generator and absorber.

The key chiller performance equation is derived from statements of the First and Second Laws specifically applied to the 4-reservoir absorption system (condenser, evaporator, generator and absorber), and of energy balance at the 4 heat exchangers. The First Law is expressed as

$$\Delta E = 0 = Q_{gen} - Q_{abs} - Q_{cond} + Q_{evap}. \tag{10.17}$$

For the expression of the Second Law, we adapt Equation (10.2) to the absorption machine, with q_{abs} and q_{gen} denoting heat transfer deriving from dissipation at the absorber and generator:

$$\Delta S = 0 = \frac{Q_{cond} - q_{cond}}{T_{cond}} + \frac{Q_{abs} - q_{abs}}{T_{abs}} - \frac{Q_{gen} - q_{gen}}{T_{gen}} - \frac{Q_{evap} - q_{evap}}{T_{evap}}. \tag{10.18}$$

In Equation (10.18), all the temperatures are process-average refrigerant temperatures. Denoting the total dissipative heat transfer associated with heat rejection as $q_{rej} = q_{cond} + q_{abs}$, and introducing the heat exchanger heat balance equations (analogous to Equations (10.4) and (10.5)), we can write the analog of Equation (10.3) for the absorption machine as:

$$\begin{aligned}\frac{1}{COP} &= \left[\frac{T_{cond}^{in} - T_{evap}^{out}}{T_{evap}^{out}}\right]\left[\frac{T_{gen}^{in}}{T_{gen}^{in} - T_{cond}^{in}}\right] \\ &+ \left[\frac{1}{Q_{evap}}\right]\left[\frac{T_{gen}^{in}}{T_{gen}^{in} - T_{cond}^{in}}\right]\left[q_{rej} - \frac{T_{cond}^{in}\, q_{evap}}{T_{evap}^{out}} - \frac{T_{cond}^{in}\, q_{gen}}{T_{gen}^{in}}\right].\end{aligned} \tag{10.19}$$

Now we invoke the additional approximation that the dominant internal irreversibilities lie in the generator and heat rejection branches. In the second term on the right-hand side of Equation (10.19), the q_{evap} term is treated as negligible, and the other loss terms, q_{rej} and q_{gen}, are viewed as constants characteristic of a particular absorption chiller. The final approximate formula for COP is then

$$\begin{aligned}\frac{1}{COP} &= \left[\frac{T_{cond}^{in} - T_{evap}^{out}}{T_{evap}^{out}}\right]\left[\frac{T_{gen}^{in}}{T_{gen}^{in} - T_{cond}^{in}}\right] \\ &+ \left[\frac{1}{Q_{evap}}\right]\left[\frac{T_{gen}^{in}}{T_{gen}^{in} - T_{cond}^{in}}\right]\left[B_1 - \frac{B_2 T_{cond}^{in}}{T_{gen}^{in}}\right]\end{aligned} \tag{10.20}$$

where the constants B_1 and B_2 characterize the irreversibilities of a particular chiller.

E3. Comparing model predictions against experimental data

To illustrate the comparison of model predictions against actual experimental data, we present results based on experimental measurements we performed on a nominal 7 kW commercial LiBr-water hot-water-fired single-stage absorption chiller [Yazaki 1979]. The cooling rate was varied by a factor of 6.5 by varying T^{in}_{cond}, T^{out}_{evap} and T^{in}_{gen} within manufacturer-prescribed ranges, as summarized in Table 10.4.

According to Equation (10.20), a plot of

$$\left[\frac{T^{in}_{gen}-T^{in}_{cond}}{T^{in}_{gen}\,\mathrm{COP}}-\frac{T^{in}_{cond}-T^{out}_{evap}}{T^{out}_{evap}}\right]Q_{evap} \text{ against } \frac{T^{in}_{cond}}{T^{in}_{gen}} \qquad (10.21)$$

should yield a straight line of slope $-B_2$ and extrapolated ordinate intercept B_1. The same straight line should represent the data for all values of T^{in}_{cond}. Experimental confirmation is illustrated in Figure 10.8, where, to within the experimental uncertainty of ±5% in the ordinate values, all 50 data points appear to be well correlated by the simple linear relation predicted (goodness-of-fit $R = 0.993$). With the regression parameters $B_1 = 21.6$ kW and $B_2 = 24.3$ kW obtained from Figure 10.8, the experimental COP values are then correlated with a rms error of 3.85% (see Figure 10.9), which is below the experimental uncertainty of ±6.2%.

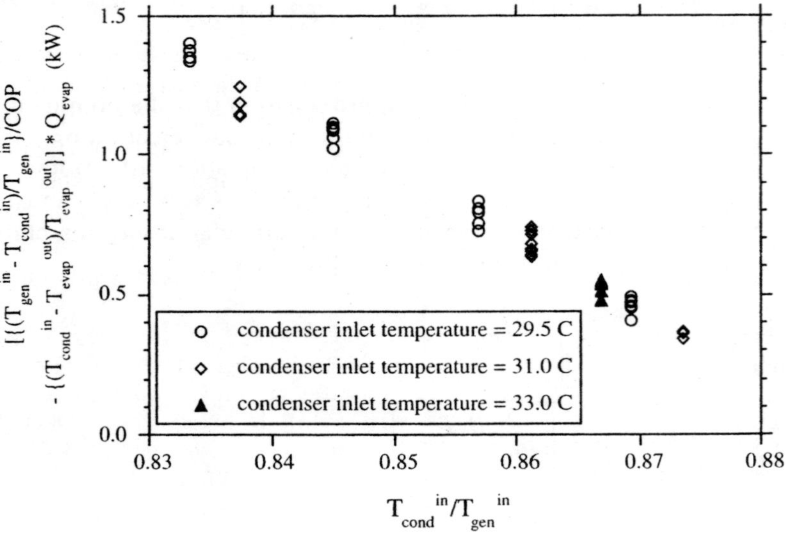

Fig.10.8 Test of the predicted functional dependence of Equation (10.20) cast as (10.21). Measurements were performed for 3 values of T^{in}_{cond}. Experimental uncertainty in the ordinate values is ±5%.

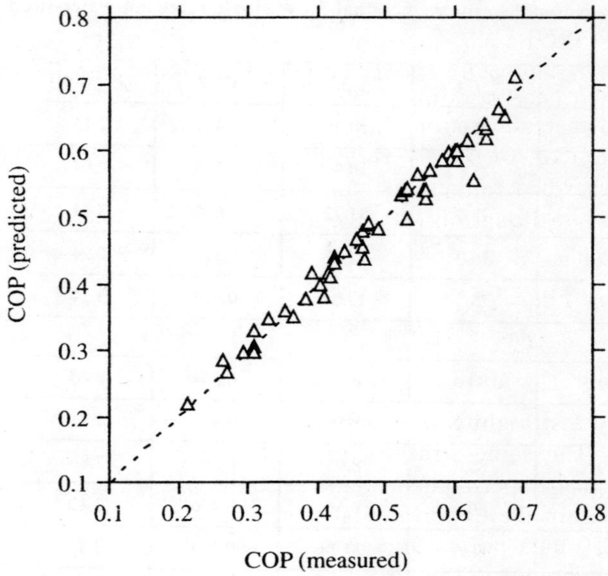

Fig.10.9 Predicted vs experimentally-measured COP for the absorption chiller. The rms error for the 50 data points is 3.85%. Experimental uncertainty in measured COP is ±6.2%.

E4. Case study on the effect of surfactant

As an additional data-based example for the value of the analytic models for single-stage absorption chillers, we cite a case study that investigated the potential improvements in COP associated with introducing a surfactant into the absorber [Ng *et al* 1999]. We'll use this experimental study to highlight two points: (a) the apparent constancy of the rate of internal dissipation in absorption chillers, *i.e.*, a *linear* characteristic plot of 1/COP against 1/(cooling rate); and (b) the clear identification of anticipated physical effects in the chiller with shifts on the characteristic plot.

First, 8 sets of measurements were performed on a hot water-fired single-stage LiBr-water absorption chiller. Its nominal rated cooling power was 5.3 kW. The condenser and evaporator coolant temperatures were fixed at T^{in}_{cond} = 31°C and T^{out}_{evap} = 8°C, while T^{in}_{gen} was varied from 61 to 95°C. The variation of COP with cooling rate is plotted in Figure 10.10 (labeled "without surfactant"). Only 7 of the 8 measured points are graphed in Figure 10.10 because the eighth point falls at such a high value of $1/Q_{evap}$ (6.6 kW^{-1}) that little detail for the remaining points would be discernible. Nonetheless, even at the anomalously low value of 3% of the rated cooling power for the ungraphed point, the slope

Table 10.4 Measured data for the nominal 7 kW single-stage hot-water-fired absorption chiller [Yazaki 1979].

T_{cond}^{in} (°C)	T_{gen}^{in} (°C)	T_{evap}^{out} (°C)	COP	Q_{evap} (kW)
29.5	75.0	10.46	0.467	2.15
29.5	75.0	10.96	0.548	2.84
29.5	75.0	11.72	0.602	3.04
29.5	75.0	12.88	0.593	2.94
29.5	75.0	13.68	0.672	3.24
29.5	80.0	8.38	0.354	2.40
29.5	80.0	8.56	0.405	2.71
29.5	80.0	8.95	0.393	2.86
29.5	80.0	9.56	0.466	3.40
29.5	80.0	10.24	0.523	3.85
29.5	80.0	10.90	0.595	4.30
29.5	80.0	11.73	0.646	4.56
29.5	80.0	12.68	0.646	4.63
29.5	85.0	9.04	0.458	4.12
29.5	85.0	9.39	0.476	4.33
29.5	85.0	9.88	0.475	4.34
29.5	85.0	10.00	0.530	4.87
29.5	85.0	10.53	0.557	4.83
29.5	85.0	10.53	0.557	4.83
29.5	85.0	11.51	0.665	6.24
29.5	85.0	13.27	0.687	6.58
29.5	90.0	7.93	0.307	2.89
29.5	90.0	10.03	0.293	2.75
29.5	90.0	11.43	0.382	3.58
29.5	90.0	13.19	0.415	3.92
29.5	90.0	15.06	0.426	4.10

Table 10.4 continued

31.0	75.0	10.82	0.419	1.56
31.0	75.0	11.46	0.559	2.15
31.0	75.0	12.87	0.584	2.41
31.0	75.0	13.37	0.619	2.41
31.0	75.0	14.41	0.618	2.49
31.0	75.0	16.12	0.644	2.51
31.0	80.0	8.81	0.264	1.65
31.0	80.0	9.10	0.307	1.95
31.0	80.0	9.51	0.329	2.07
31.0	80.0	10.06	0.425	2.70
31.0	80.0	10.77	0.491	3.10
31.0	80.0	11.45	0.559	3.55
31.0	80.0	12.19	0.605	3.91
31.0	80.0	13.15	0.603	3.96
31.0	90.0	8.19	0.269	2.38
31.0	90.0	9.69	0.366	3.22
31.0	90.0	11.03	0.470	4.14
31.0	90.0	12.6	0.531	4.74
33.0	80.0	10.27	0.213	1.01
33.0	80.0	10.96	0.307	1.44
33.0	80.0	11.61	0.411	1.93
33.0	80.0	12.29	0.440	2.38
33.0	80.0	12.81	0.532	3.04
33.0	80.0	13.67	0.564	3.20

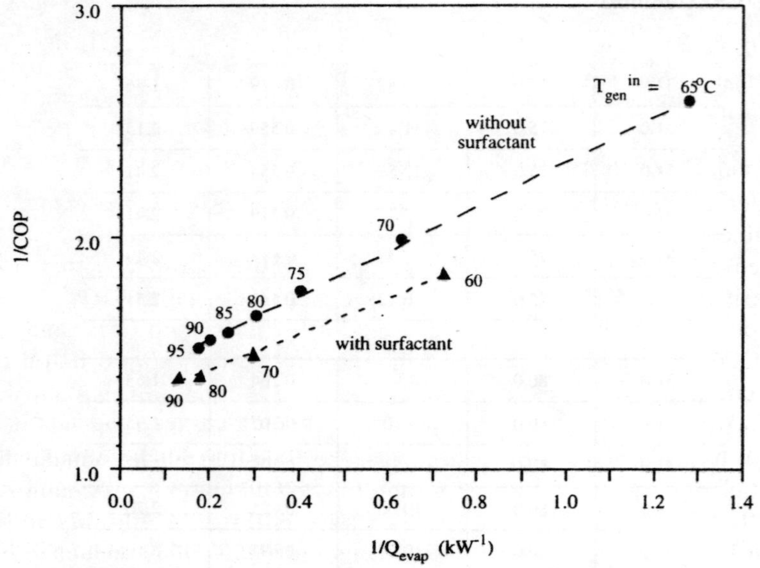

Fig.10.10 Characteristic plot for the nominal 5.3 kW single-stage LiBr-water absorption chiller, with and without surfactant added to the absorber. T_{cond}^{in} and T_{evap}^{in} are fixed for all measurements at 31°C and 8°C, respectively. Values of T_{gen}^{in} are noted for each point. For a fixed value of T_{gen}^{in}, there is a marked increase in cooling rate when surfactant is introduced. The broken lines are linear regression best fits. After the addition of surfactant, both the slope and ordinate intercept decrease, as expected.

and intercept of the linear regression remain unaffected.

As a means of improving chiller COP, a small volume (less than 1%) of the surfactant ethyl-hexanol was then added to the absorber. Surfactant reduces the surface tension of the LiBr-water solution at both the absorber and the generator. The anticipated effects are:

(1) reduced mass-transfer resistance in the absorber and generator, and hence a lowering of the slope of the characteristic plot;

(2) an improvement in heat transfer in the absorber and generator, which should be manifested as a decrease in the ordinate intercept in the characteristic plot; and

(3) both of these effects should give rise to a noticeable rise in cooling rate, for a given value of T_{gen}^{in}, when surfactant is present.

In addition, if the rate of internal dissipation is roughly constant and dominant (as in other single-stage absorption chillers examined), the characteristic plot should be well approximated as a straight line.

Four sets of measurements were carried out on the chiller *with* surfactant, at the same condenser and evaporator coolant temperatures, and with T_{gen}^{in} spanning 60–90°C. These 4 points are also plotted in Figure 10.10. Linear regression best-fits are included. Commensurate with the physical picture underlying the analytic chiller model, all 3 effects noted above are observed.

E5. The extended performance curve

Chiller operating parameters in this case place chiller performance on the linear part of the characteristic curve. Absorption chillers can be driven at sufficiently high relative cooling rates, however, that non-linearities become noticeable (in a plot of 1/COP against $1/Q_{evap}$). One can observe chiller operation at the point of maximum COP, and can see COP decreasing on the *high* cooling rate side, as well as on the low cooling rate side.

To illustrate this point, we cite performance curves (and not individual measurements) for steam- and hot-water fired commercial single-stage absorption chillers as presented in a manufacturer's catalog [Trane 1989]. One set of curves, presented as relative generator heat input against relative cooling rate, pertains to 23 different single-stage chillers that range in size from 355 to 5840 kW. The term "relative" is used because for each chiller, both the generator heat input and the cooling rate are expressed relative to their values at nominal design-load conditions.

Figure 10.11 presents these curves plotted as the ratio $\dfrac{\text{COP}_{\text{design load}}}{\text{COP}}$ against the ratio $\dfrac{\text{cooling rate}_{\text{design load}}}{\text{cooling rate}}$ at 5 values of T_{cond}^{in}. Essentially, this is a plot of $\dfrac{\text{constant}}{\text{COP}}$ against $\dfrac{\text{constant}}{\text{cooling rate}}$ where the constants are different for each of the 23 chillers. The curves in Figure 10.11 reveal that 1/COP is indeed linear in 1/(cooling rate), as per the model prediction, for almost the entire range of operating conditions.

At the higher cooling rate end, these chillers exhibit a maximum COP value, beyond which there is a narrow region where COP decreases as cooling rate increases further. A more fundamental treatment of this performance regime was presented in Chapters 4 and 9.

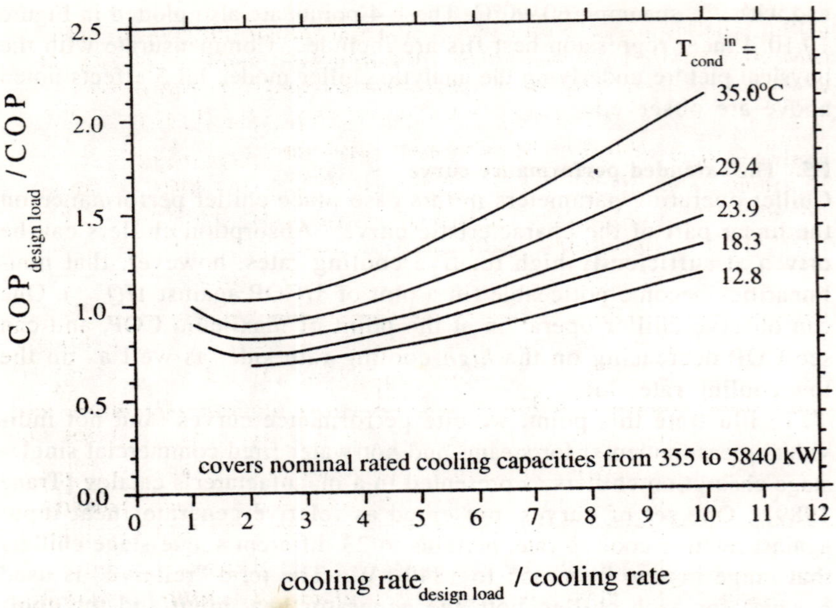

Fig.10.11 Normalized absorption chiller performance curves for 23 different single-stage machines, taken from [Trane 1989].

F. LESS CONVENTIONAL CHILLERS: THERMOACOUSTIC AND THERMOELECTRIC REFRIGERATORS

F1. Background

This section deals with two less conventional chillers: thermoacoustic and thermoelectric refrigerators. The basic principles of their operation were reviewed in Chapter 2. We will not address accurate predictive and diagnostic tools here: in the case of the thermoacoustic chiller due to the dearth of published experimental data available to date as well as the complexity of the problem, and in the case of the thermoelectric refrigerator, due to its being essentially a fully-solved problem for which the quasi-empirical model covers only a limited regime of operating conditions. In addition, irreversibilities in the thermoelectric chiller are different from those in mechanical and absorption chillers. Therefore we aim simply to examine the principal trends in chiller performance, and to compare them against the predictions of the simple thermodynamic models developed here.

F2. Thermoacoustic chillers

Recall the schematic of the thermoacoustic chiller (Figure 2.24). The thermoacoustic refrigerator has external irreversibilities such as finite-rate heat transfer at the heat exchangers, and internal irreversibilities such as fluid friction and imperfect thermal contact between the acoustically oscillating working fluid and the stack plates. Figure 10.12 is based on published experimental data for a small laboratory thermoacoustic chiller [Garret & Hofler 1992]. Computer simulation results were not considered for this analysis - only measured performance.

Figure 10.12 shows that the qualitative behavior of the thermoacoustic

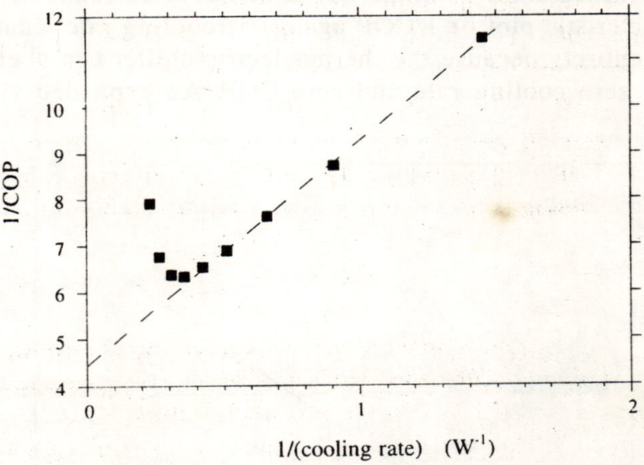

Fig.10.12 Characteristic chiller plot from measured data for a small thermoacoustic refrigerator [Garret & Hofler 1992].

refrigerator is consistent with the behavior predicted by the thermodynamic models. A broad range of cooling rates and COPs is covered. There is a linear region at cooling rates low enough for internal dissipation to dominate, and a high cooling rate regime dominated by external losses. In the intermediate region is a point of maximum COP.

F3. Thermoelectric chillers

Recall the schematic of the thermoelectric chiller (Figure 2.25) and the governing performance equations (Equations (2.20)–(2.23)). The thermoelectric refrigerator is essentially a solved problem in terms of analytic modeling equations. Hence there is no empiricism in predicting or analyzing the characteristic performance curve once the material properties are known.

High-efficiency thermoelectric materials inherently possess both a high thermal conductivity and a high electrical conductivity. Unlike almost all other chillers, finite-rate heat exchange losses are negligible (relative to the other loss mechanisms). The two key sources of irreversibility are electrical resistance and heat leak. The heat leak militates against slow operation, *i.e.*, low electrical current, and electrical resistance mitigates against fast operation, *i.e.*, high current. Therefore we expect a plot of COP against cooling rate to be a loop diagram: cooling rate (and hence COP) vanishes in two distinct limits of low and high current (the roots of Equation (2.20)). Both COP and cooling rate exhibit (measurable) maxima, but at different currents.

The characteristic plot of 1/COP against 1/(cooling rate) cannot be drawn in its entirety because the thermoelectric chiller can operate to the limits of zero cooling rate and zero COP. An expanded view of

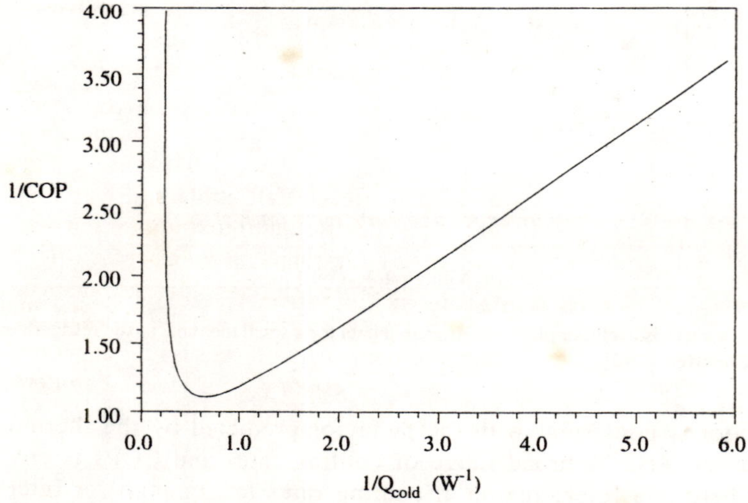

Fig.10.13 Characteristic chiller plot for a commercial thermoelectric refrigerator with T_{cold} = 5°C and T_{hot} = 35°C. The material properties are: α = 0.0008 V K^{-1}, R_{el} = 0.004 ohm, and K = 0.06 W K^{-1}. An expanded view is shown because both COP and cooling rate can be brought to zero in normal operation.

the characteristic plot is shown in Figure 10.13 for a commercial thermoelectric refrigerator that can provide up to about 4.5 W of cooling power. Heat conduction (heat leak loss) governs the seeming linear dependence of 1/COP on $1/Q_{cold}$ at low cooling rates. Figure 10.14 is

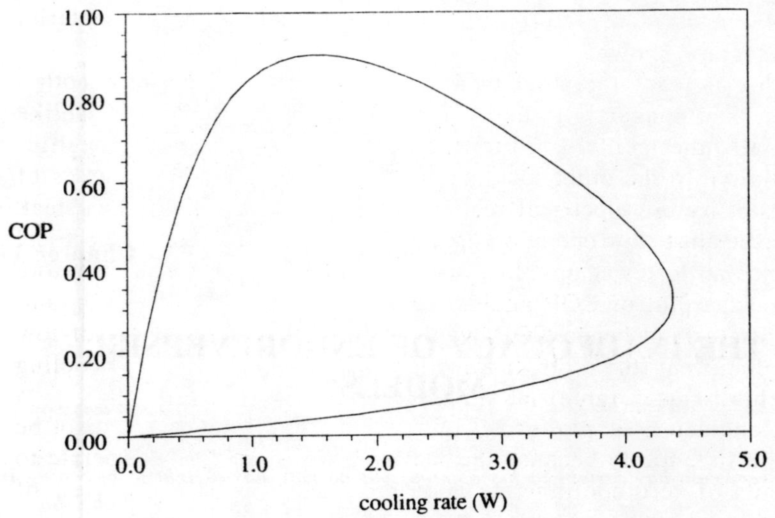

Fig.10.14 Performance characteristic of the same thermoelectric refrigerator plotted so as to show the full operating range.

a plot of COP against cooling rate for the same device, which shows the full range of operating conditions.

F4. Unique thermodynamic aspects of thermoelectric chillers

An interesting contrast between thermoelectric and most common commercial chillers is the region of the characteristic plot where chillers are operated. As noted in earlier chapters, mechanical and absorption chillers are operated in the linear regime of the 1/COP *vs* 1/(cooling rate) plot, and usually not beyond the cooling rate at which COP peaks. For economic and mechanical reasons, conventional chillers are purposely designed to restrict their operation to only part of the theoretically-possible range.

Thermoelectric chillers, however, are operable over the full range of theoretical cooling rates. For economic reasons, commercial thermoelectric devices are usually operated between the points of maximum COP and maximum cooling rate, on the high-COP branch (Figure 10.14). Because finite-rate heat exchange to and from the reservoirs is not a bottleneck, thermoelectric chillers can readily realize the regime of relatively high cooling rates, as well as the "loop" regions where two different values of COP are possible for the same cooling rate.

Chapter 11

THE INADEQUACY OF ENDOREVERSIBLE MODELS

"Every man has a right to his opinion, but no man has a right to be wrong in his facts."

- Bernard Baruch

A. MISSING MOST OF THE PHYSICS AND ITS CONSEQUENCES

Internal entropy production typically dominates the thermodynamic performance of real chillers. For typical reciprocating chillers, around 60% of the total losses derive from internal dissipation. The corresponding figure for common absorption chillers is over 50%. One can analyze the measurements reported in Chapters 6-10 to confirm that the 50–60% figures are the rule rather than the exception. The quasi-empirical models of Chapter 10 are predicated on internal losses governing chiller COP. The accuracy of the model's predictive and diagnostic capabilities is difficult to account for other than by the dominance of internal losses.

As sample exercises for mechanical chillers, start with the analytic chiller performance equations of Chapter 5. Plug in realistic values of coolant temperatures, cooling rates and chiller characteristic parameters. Now examine the relative contribution of each term to COP. Internal losses are always significant, and typically the dominant factor (*e.g.*, Tutorial 6.1).

Any chiller model that summarily excludes internal dissipation clearly omit's more than half the physics of the problem; it cannot offer predictions that correspond to reality. Hence it is surprising to discover the large number of journal articles that propose endoreversible models for the behavior of real chillers. (Rather than provide an extensive list of these articles here, we refer the reader to the papers where the shortcomings of endoreversible chiller models were first documented

[Gordon & Ng 1994a; Gordon & Ng 1994b; Chua *et al* 1996; Chua *et al* 1997; Ng *et al* 1997a, Ng *et al* 1997b, Ng *et al* 1997c], and which in turn provide lengthy citation lists for the endoreversible chiller model papers.) The endoreversible model accounts solely for the irreversibility of finite-rate heat transfer in the heat exchangers. Clearly the COP of an endoreversible chiller (Equations (4.22) and (4.34)) is closer to reality than the COP of a reversible chiller (Equations (2.5) and (2.15)). In Chapters 2 and 4, we showed how the endoreversible limit emerges as a special case of general thermodynamic models when internal dissipation and heat leaks are consciously ignored.

But most endoreversible model studies purport to provide reasonable predictions of COP, to determine optimal conditions for chiller operation, and to predict the fundamental chiller performance curve. In these regards, it should be stated unequivocally that the endoreversible model yields gross errors. Specifically, endoreversible models provide patently incorrect and inaccurate predictions of: (*a*) COP values; (*b*) how COP varies with cooling rate; (*c*) the dependence of COP on the properties of the principal chiller components; and (*d*) the ability to optimize chiller design with respect to these parameters. We devote this chapter to focusing upon the fine points of a few examples analyzed in earlier chapters, and to highlighting the errors inherent to endoreversible models.

Endoreversible chiller models have been published under the heading of finite-time thermodynamics. Indeed, they invoke irreversible heat transfer, *i.e.*, requiring finite time for a given energy transfer (in contrast to the reversible limit in which infinite time is required). But the fruitful approach of finite-time thermodynamics should not be indicted because of modeling approaches that invoke it incompletely or misleadingly.

A few finite-time thermodynamic models that capture the correct *qualitative* trend of chiller performance, in terms of how COP varies with cooling rate, have been presented [Andresen *et al* 1977; Alefeld 1987; Grazzini 1993]. Although these models point in the right direction, they offer few, if any, quantitative comparisons with actual devices. An important question that has remained unanswered is to what extent these approaches truly capture the basic physics of real chillers, and whether they can be realistically related to experimental realities.

We devoted Chapter 7 to what we view as a true finite-time thermodynamic analysis of real chillers. In fact, due to the fact that time is not the only variable the availability of which should be treated as limited, we would refer to the approach as finite-resource thermodynamics. For example, finite heat exchanger inventory may be no less significant than finite time. Such a finite-resource optimization was covered in Chapters 6 and 7.

While we are not advocates of focusing upon the errors or misleading claims of others, the proliferation of journal papers that propose the endoreversible model for serious chiller analysis, including chiller optimization, prompted us to set the record straight in unambiguous terms. A particularly suspicious aspect of the endoreversible model studies is that not a single datum of actual chiller performance is offered to test the theoretical models. In light of the extensive data base available in the published literature and in readily-available manufacturer catalogs, one cannot help but wonder why comparisons against chiller data have been systematically omitted.

B. PREDICTING COP AS A FUNCTION OF COOLING RATE

The characteristic chiller performance curve, and the manner in which it is comprised of contributions from external and internal losses, was reviewed in Chapters 2 and 4-9, including the isolated endoreversible contributions. The endoreversible chiller model misses the key feature of the maximum COP point and the large penalty in COP paid for operating at relatively low cooling rates (*e.g.*, Figures 1.4, 6.3, 6.4, 9.1, 9.2, 10.7 and 12.2). Its predictions are qualitatively incorrect, in that they indicate chiller COP should uniformly decrease as cooling rate increases.

Endoreversible chiller models supplemented with heat leaks have also been proposed, their virtue being a prediction of a point of maximum COP, as opposed to the pure endoreversible model where COP increases monotonically to its maximum Carnot value in the limit of vanishing cooling rate. (For cryogenic refrigerators, heat leaks are often considerable; but even then ignoring internal dissipation translates into omitting a substantial fraction of the total dissipation.) However, these modified endoreversible models still fail to capture the principal losses associated with internal entropy production in commercial chillers, and hence offer predictions that are distant from reality. By inserting experimentally-measured heat leak and heat exchanger values, we show below that the predictions even of these heat-leak enhanced models are far different than actual chiller performance.

Are there real chillers that are well approximated by the endoreversible model? Can commercial chillers be driven such that internal dissipation becomes negligible relative to external losses? The answer is yes, it is possible, but extremely undesirable and therefore never implemented - a reality reflected in the construction, configuration and operation of real commercial mechanical and absorption chillers.

Given even state-of-the-art technology for compressors, throttlers and heat exchangers (and additionally of generators and absorbers for absorption machines), internal entropy production lowers the chiller COP far below the reversible Carnot limit. To produce a chiller with relatively negligible internal losses at acceptable cooling rates, one needs to markedly increase external losses, *e.g.*, to use poor heat exchangers. Certainly this is possible, but clearly unwanted, due to a considerable and avoidable lowering of COP.

C. ANALYSIS WITH DATA FROM RECIPROCATING CHILLERS

To sharpen these points, let's use the experimental measurements cited in Chapter 6 for two commercial reciprocating chillers. We generate the chiller curves that correspond to: (1) actual chiller performance; (2) a pure endoreversible chiller; and (3) an endoreversible chiller with

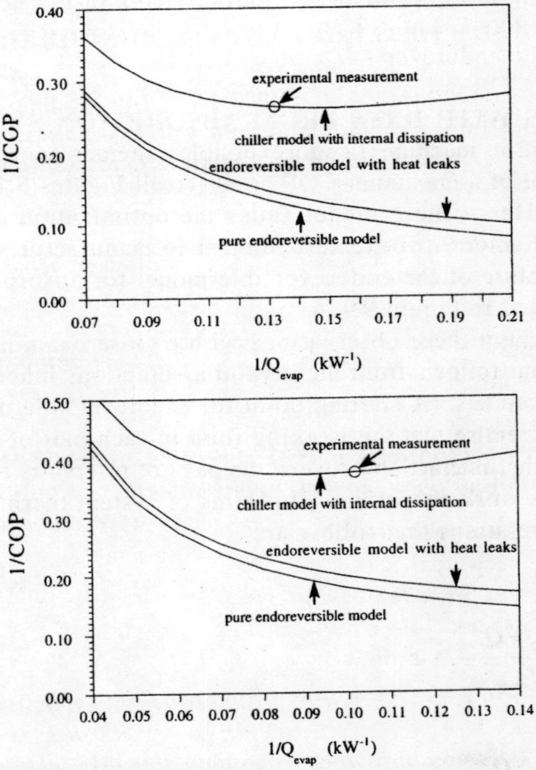

Figure 11.1: Characteristic performance curves for two water-cooled reciprocating chillers, along with curves calculated from the pure endoreversible chiller model and from an endoreversible chiller model with heat leaks. The heat exchanger and heat leak thermal conductances have been measured experimentally, *i.e.*, there are no adjustable parameters. (*a*) For the chiller reported in [Liang & Kuehn 1991]. (*b*) For the chiller reported in [Chua *et al* 1996].

heat leaks. These curves are drawn in Figures 11.1a and 11.1b. Note that these calculated curves involve no adjustable parameters. All the principal chiller parameters were measured directly.

The comparison between the data-based chiller curves and the curves based on the endoreversible models graphically indicates the dominant contribution of internal losses to chiller COP, at least for realistic chiller operating conditions. Finite-rate heat transfer by itself (the pure endoreversible model) is one important element, but is inadequate as a complete model for real chillers. In principle, the introduction of heat leaks gives rise to a maximum COP point. However, for actual heat leak values, the predicted maximum COP point is so far from the actual one, and at such high values of COP, as to render these chiller models untenable. The endoreversible COP predictions are around a factor of 2 or more larger than real chiller COPs in the actual operating range.

D. ANALYSIS WITH DATA FROM ABSORPTION SYSTEMS

For real absorption machines, endoreversible schemes cannot account for the existence of a maximum-COP point (recall Figures 5.1, 9.1, 9.2, 9.3 and 9.4). Hence these models miss the optimization capability, portrayed in Chapter 9, of greatest interest to manufacturers and designers. The failure of the endoreversible model for absorption chillers is exemplified in Figure 9.1.

Let's supplement these observations with a close examination of a subtle failure that follows from the invalid assumptions inherent in the endoreversible models. A starting point for endoreversible model predictions is the premise that the working fluid in each pair of reservoirs is isolated, and that interaction between the pairs of reservoirs is achieved via entropy-less work conversion. In terms of system thermodynamic variables, two relations that follow are

$$\varepsilon_1 = \frac{Q_{\text{cond}}}{T_{\text{cond}}} - \frac{Q_{\text{evap}} + Q_{\text{evap}}^{\text{leak}}}{T_{\text{evap}}} \geq 0 \qquad (11.1)$$

$$\varepsilon_2 = \frac{Q_{\text{cond}}}{T_{\text{cond}}} - \frac{Q_{\text{gen}} + Q_{\text{gen}}^{\text{leak}}}{T_{\text{gen}}} \geq 0. \qquad (11.2)$$

(To be as liberal as possible in evaluating endoreversible models, we have retained the heat leak terms in Equations (11.1) and (11.2). In reality, however, the refrigerant is in thermal communication with all 4 heat reservoirs.)

The modeling assumptions can be checked against the experimental measurements and validated against the computer simulation results cited in Chapter 9 for absorption chillers and heat pumps. So we calculated ε_1 and ε_2 from actual performance data and list them in Table 11.1. Note that relations (11.1) and (11.2) are almost universally violated. No fundamental physical law is at stake here. The only relation that is constrained by the Second Law is the non-negativity of the internal entropy production.

For the absorption heat transformer, the corresponding endoreversible modeling assumptions translate to

Table 11.1: Check of the consistency of fundamental predictions of endoreversible models against experimental and simulated results for absorption chillers, heat pumps and heat transformers. The endoreversible model requires that the variables ε_1, ε_2, ε_3 and ε_4 defined in (11.1)-(11.4) be non-negative. This requirement is almost universally contradicted by the data.

Absorption machine	ε_1 (kW K^{-1})	ε_2 (kW K^{-1})	ε_3 (kW K^{-1})	ε_4 (kW K^{-1})
single-stage chiller [Carrier 1962]	–0.247	–0.559		
single-stage chiller [Chuangh & Ishida 1990]	–1.94	–0.811		
double-stage series-cycle chiller [Chuang & Ishida 1990]	–7.91	–0.811		
double-stage parallel-cycle chiller [Chuang & Ishida 1990]	–7.44	0.548		
single-stage heat pump [Abrahamsson *et al* 1995]	–0.119	–0.433		
single-stage heat transformer [Abrahamsson *et al* 1995]			–0.00112	–0.00191
single-stage heat transformer [Zhou & Machielsen 1996]			–0.352	–0.850
double-stage heat transformer [Zhou & Machielsen 1996]			–3.75	–1.43
triple-stage heat transformer [Zhou & Machielsen 1996]			–7.23	–2.00
compact triple-stage heat transformer [Zhou & Machielsen 1996]			–4.77	–7.27

$$\varepsilon_3 = \frac{Q_{abs} + Q_{abs}^{leak}}{T_{abs}} - \frac{Q_{gen} - Q_{gen}^{leak}}{T_{gen}} \geq 0 \qquad (11.3)$$

$$\varepsilon_4 = \frac{Q_{abs} + Q_{abs}^{leak}}{T_{abs}} - \frac{Q_{evap} - Q_{evap}^{leak}}{T_{evap}} \geq 0. \qquad (11.4)$$

The values of ε_3 and ε_4 are also included in Table 11.1. Every heat transformer considered violates relations (11.3) and (11.4).

The combined experimental evidence would appear to render endoreversible models for absorption chillers, heat pumps and heat transformers untenable, be it in accounting for qualitative trends or in quantitative predictions of system performance.

E. ARE ENDOREVERSIBLE MODELS FOR HEAT ENGINES ANY BETTER?

Even more attention has been devoted in the professional literature to endoreversible models for heat engines than for chillers. Whereas chillers are characterized by their COP-cooling rate relation, heat engines can be characterized by their power-efficiency graphs. The bounds on heat engine efficiency (defined as the work generated divided by the heat input) that derive from endoreversible models are certainly closer to reality than the limiting reversible or Carnot efficiency. But we must ask if, as we found for chillers, those endoreversible bounds are so far from reality as to render them intellectual curiosities. An important basis for this question is the observation that the measured efficiencies of several power plants are close to those predicted from endoreversible models.

Rather than reproduce published results, we simply cite them here and refer the reader to the article in which the calculations and experimental results are reported [Gordon & Huleihil 1992]. And as explained above specifically for chillers, finite-time thermodynamic modeling of heat engines is not the issue here. Rather what we are challenging is restricting that modeling to external losses while completely excluding factors that turn out to dominate heat engine performance: internal losses and limitations related to the equation of state of the working fluid.

In [Gordon & Huleihil 1992], several key points are established and confirmed with data from real heat engines.

1) The actual efficiencies of gas-cycle and steam-cycle power plants are primarily dictated by internal losses from fluid and mechanical friction. Endoreversible models offer poor quantitative predictions for a

variety of real heat engines.

2) The agreement noted between endoreversible efficiencies and the efficiency of a couple of real power plants turns out to be fortuitous. Endoreversible models are shown to noticeably overpredict heat engine efficiencies.

3) The difference between the maximum-efficiency and maximum-power points of heat engines typically is dramatically less than that predicted by endoreversible engine models.

4) The thermoelectric generator is a real example where heat exchange losses are negligible. Thermoelectric performance is completely determined by internal dissipation and heat leaks (which is clear from the energy balance equations cited in Chapter 2, but adapted to heat engines rather than chillers). Hence the endoreversible model misses essentially 100% of the physics of the problem.

Chapter 12

HEAT EXCHANGER INTERNAL DISSIPATION IN CHILLER ANALYSIS AND THE ESSENTIAL ROLE OF ACCURATE PROCESS AVERAGE TEMPERATURES

"That is the essence of science: ask an impertinent question, and you are on the way to a pertinent answer." - Jacob Bronowski

A. PEEKING INTO THE BLACKBOX

In their simplest and most approximate form, the analytic chiller models developed in Chapters 4–5 and in Chapter 10 allow us to relate to a cooling device as a blackbox. By performing a number of completely external measurements – the non-intrusive type commonly provided in manufacturer catalogs and easily conducted in-house – we can use simple regression techniques to characterize a chiller by three parameters. We needed a basic notion of what transpires inside the blackbox, but we did not require any internal measurements. We discovered that simple chiller performance formulae offered predictive tools with some diagnostic capabilities. Some of the studies covered in earlier chapters are in this spirit.

In order to develop a more comprehensive diagnostic tool, and in order to perform optimization studies, we need to peek inside that blackbox and obtain additional information about the internal components – additional measurements on the heat exchangers and on the refrigerant as it enters or leaves the different chiller compartments. For simplicity, we aspire to keep the additional required information to a minimum.

In this chapter, we'll open the chiller, perform a few key measurements that are suggested by the fundamental model of Chapter 5, and see how to exploit that model to diagnose and optimize chillers. The fundamental model in its extended form – Equation (5.3) – indicates that at the very least we also need:

(1) To carefully distinguish among external, internal and heat leak losses.

(2) To be careful that internal losses are not counted as external losses. For example, pressure drops in the heat exchangers, which are internal dissipation, should not enter the bookkeeping as part of the finite-rate heat exchange irreversibility. If these internal heat exchanger losses are insignificant, then clearly this is a mute point; but we'll see that often this is not the case.

(3) To measure how the external losses are partitioned between the evaporator and the condenser.

At the heart of these procedures lies the accurate determination of the process average temperatures (PATs). First, we'll see why accurate PATs are crucial for establishing how losses are divided not only among internal, external and heat leak losses, but also among the different sources of internal losses. A quantitative determination of how losses divide among the key chiller components is itself an important diagnostic step.

Second, we'll challenge the conventional wisdom that heat exchanger internal losses are insignificant relative to the other internal losses. With experimental data from real chillers, we'll show that overlooking internal losses in the heat exchangers can lead to non-negligible errors in diagnostic studies.

And third, optimization case studies will be presented to highlight how the precision analysis of internal losses and PATs translates into information of considerable importance to designers and manufacturers. One intriguing observation – a reaffirmation of findings in earlier chapters – is the extent to which commercial chillers have empirically evolved to optimal performance configurations for given technological constraints.

Experimental measurements of the precision and scope required to carry out these calculations are not commonly available. For example, they do not appear in manufacturer catalogs or most journal articles on these topics. Therefore, in order to investigate these effects and to arrive at meaningful conclusions, we needed both to perform our own measurements, and to tap the few studies available from manufacturers and in the journal literature that allow such detailed assessments to be made.

Section B is a case study for a reciprocating chiller [Ng *et al* 1998a]. The specific implications of determining accurate PATs that we'll explore are: (*a*) diagnostics for the heat exchangers, with the ability to distinguish clearly between the evaporator and the condenser; and (*b*) accurate determination of the chiller characteristic performance curve for the identification of conditions at which COP is maximized.

In Section C, a corresponding study is presented for the more complex case of an absorption chiller – more complex in the sense that more control variables are involved [Ng *et al* 1998b]. The implications we'll examine for chiller diagnostics and optimization are similar to those in Section B.

B. STUDIES FOR A RECIPROCATING CHILLER

B1. Background to the problem

As we have seen in Chapters 6 and 10, one clear message that emerges from the analysis of chiller data is that the thermodynamic performance of real chillers is dominated by internal dissipation. In performing the entropic bookkeeping, most researchers who have recognized the key role of internal losses have restricted entropy production measurements or calculations to the compressor and the expansion device. The conventional wisdom has been that internal entropy production in the chiller's heat exchangers is negligible.

Hence the PATs used in those analyses were calculated without accounting for the internal dissipation in the evaporator and condenser heat exchangers. As we'll see shortly from careful experimental measurements and basic thermodynamic analysis, ignoring internal entropy production in the heat exchangers can introduce noticeable errors in predicting and diagnosing chiller behavior, as well as sometimes leading to unphysical PATs.

B2. Experimental details and thermodynamic calculations

The experimental measurements used here were originally reported for a water-cooled vapor-compression reciprocating chiller in [Chua 1995; Chua *et al* 1996; Gordon *et al* 1997] and were reviewed in Chapter 6. The chiller, with refrigerant Freon R12, includes a 5.6 kW semi-hermetic refrigerant-cooled compressor and coaxial heat exchangers. The chiller rated conditions are T_{evap}^{in} = 12.4°C, T_{cond}^{in} = 29.0°C, and Q_{evap} = 10.66 kW, at which point COP = 2.68.

Figure 12.1 is a *T–S* plot for the cycle – simply Figure 4.2 enhanced with indicators for the assorted PATs. The measured pressures and temperatures of the refrigerant R12 for steady-state cyclic operation at the rated operating conditions were listed in Table 4.1, along with the computed thermodynamic states at each point along the cycle. In Figure 12.1 and Table 4.1, state points 1-2-3-4-5-6 are determined based on

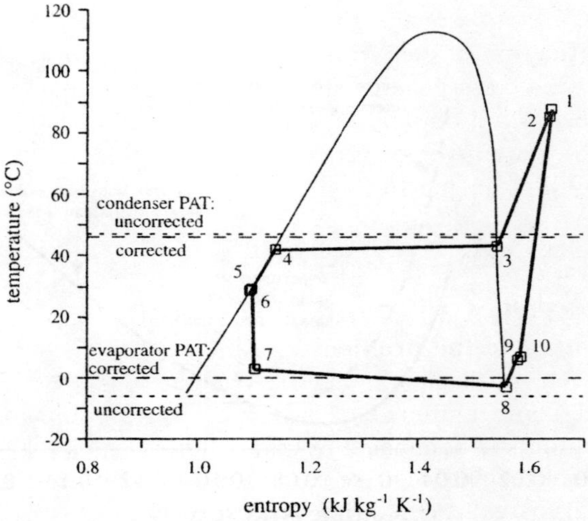

Figure 12.1: Temperature–entropy diagram for the chiller cycle at standard rated conditions, including state points 1-10. The PATs before and after incorporation of heat exchanger dissipation are also indicated. Note that the uncorrected evaporator PAT is outside the range of measured temperatures.

the temperature and pressure measurements at the inlet and outlet of the condenser, while state points 7-8-9-10 are the corresponding states in the evaporator.

For each heat exchanger, the pressure and temperature of the refrigerant at the inlet and exit were measured. The associated change in entropy and enthalpy were then computed from standard tables of the refrigerant's thermodynamic properties [Mayhew & Rogers 1971, ASHRAE 1998]. The internal entropy production in the heat exchangers follows from Equations (4.2) and (4.3). The data permit the evaluation of the entropy production from each separate chiller component. These evaluations follow the procedures illustrated in Tutorial 4.1.

By forcing the chiller performance formula, Equation (5.3), to pass through the experimentally measured point at the nominal rated conditions, we can plot the characteristic chiller curve, both with and without a proper accounting for heat exchanger internal losses. The results are drawn in Figure 12.2. The division of internal and total entropy production among the principal components is summarized in Table 12.1.

B3. Observations about internal dissipation

Several points merit particular note.

(1) Internal entropy production is about 60% of the total chiller

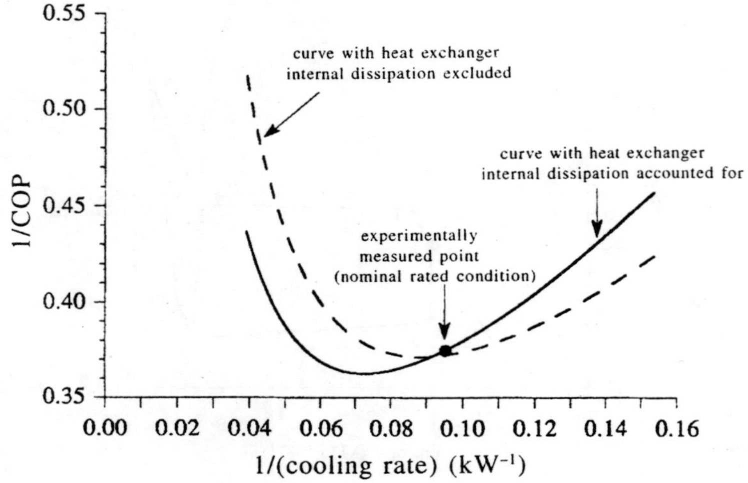

Figure 12.2: Chiller characteristic performance plots when heat exchanger internal dissipation is excluded (broken curve) and accounted for properly (solid curve).

Table 12.1: Measured *relative* contribution of each chiller component to the internal entropy production ΔS_{int} and to the total (internal plus external) entropy production in the chiller ΔS_{total}. The experimental uncertainty is ±0.02. $T^{in}_{evap} = 12.4°C$. T^{in}_{cond} spans the range 23.8–35.0°C.

ΔS_{int}				
condenser	throttler	evaporator	compressor	
0.02	0.07	0.13	0.78	
ΔS_{total}				
internal	evaporator (finite–rate heat exchange)	evaporator (heat leak)	condenser (finite–rate heat exchange)	condenser (heat leak)
0.60	0.11	0.09	0.19	0.01

entropy production. The remaining 40% is comprised of around 30% in the finite-rate heat exchange irreversibility, and the remaining 10% in heat leaks between chiller components and their environment.

(2) Of the internal entropy production, 78% derives from the compressor and 7% from the throttler.

(3) The remaining 15% of the internal entropy production, due to the pressure drop in the heat exchangers, is comprised of 13% in the evaporator and 2% in the condenser. It is this loss that has been omitted from earlier chiller analyses. A thermodynamic model that aspires to predict chiller performance yet ignores internal dissipation in the heat exchangers would introduce a non-negligible error in accounting for the curve of COP versus cooling rate, and in being of value for chiller diagnostics where the essential point is quantitative identification of the assorted sources of entropy generation.

B4. Repercussions for diagnostics and optimization

Say we wish to determine the effective thermal conductance mCE of a chiller heat exchanger with non-intrusive measurements only. The diagnostic value could be to compare the measured mCE against the manufacturer's rating, or to monitor whether heat exchanger effectiveness (and hence chiller COP) is degrading with time in an installed system. We would invoke the equation for heat transfer rate Q

$$Q = mCE\ (\text{PAT} - T^{\text{in}}) \tag{12.1}$$

where T^{in} is the coolant inlet temperature. E is readily calculated since Q, T^{in} and m are measured, and C is known.

An inaccurate expression for the PAT yields misleading values for E. Even an error of only one degree in the PAT can give rise to an observable error in the determination of E, especially for the evaporator heat exchanger, where the temperature difference, PAT $- T^{\text{in}}$, is only a few degrees. Namely, the issue is not necessarily the absolute magnitude of the error in the PAT, but its magnitude relative to the heat exchanger temperature difference.

To strengthen this point with experimental data, we present in Table 12.2 the values of heat exchanger mCE first calculated according to earlier prescriptions in which heat exchanger internal dissipation is ignored, and then calculated correctly with the measured internal entropy production accounted for. The error in the evaporator mCE is greater than that for the condenser because the evaporator incurs a higher pressure drop as a consequence of boiling being a more turbulent process than condensation.

Only with a proper PAT that accounts for heat exchanger internal dissipation does one arrive at both the correct mCE values and the correct ΔS_{int}. The maximum COP value may not change significantly as a consequence of these inaccuracies. Yet the sensitivity of COP to cooling

Table 12.2: Illustration of the impact of proper accounting of internal dissipation in the heat exchangers by comparing the computed thermal conductances (*mCE*), internal entropy generation (ΔS_{int}), and maximum COP for the chiller analyzed, with and without heat exchanger internal entropy production in the calculation. Calculations were performed at the chiller rated conditions.

	excluding heat exchanger dissipation			including heat exchange dissipation		
	condenser	evaporator	compressor + throttler	condenser	evaporator	compressor + throttler
mCE kW K^{-1}	0.838	0.593	----	0.880	0.869	----
ΔS_{int} kW K^{-1}	0	0	0.005113	0.000134	0.000911	0.005113
minimum 1/COP	0.369			0.362		

rate and coolant temperatures, and hence the determination of optimal chiller operating conditions, do change (see Figure 12.2).

In this procedure of parameter identification for the chiller, experimental measurements serve as the input data. Therefore the total dissipation will always include heat exchanger internal dissipation. However a key issue for diagnostic and optimization procedures is the division of the total dissipation among internal losses, heat leaks and external losses. Exactly how these 3 sources of irreversibility are partitioned impacts the chiller performance curve of Figure 12.2.

For chiller heat exchangers, exact measurements of refrigerant temperature at every point along the heat exchangers are not feasible. These refrigerant temperatures have a considerable spatial dependence due to processes such as de-superheating and pressure drop. By adopting the PAT approach, we can account for the influence of this non-isothermal characteristic on entropy flows. In turn, this impacts the correct identification of heat exchanger thermal conductances and internal dissipation. The PAT is not a panacea for analyzing chiller thermal dynamics; rather it is a convenient and powerful tool for the evaluation of assorted sources of dissipation and how they contribute quantitatively to chiller COP.

C. STUDY FOR AN ABSORPTION CHILLER

C1. The nature of the study

Absorption chillers also operate well below their reversible or even endoreversible limits because their thermodynamic behavior is commonly dominated by internal dissipation. In this section, we'll be showing that a significant part of that internal dissipation occurs in the heat exchangers – a key missing element in previous modeling efforts.

This omission translates into an inaccurate PAT which in turn leads to deducing inaccurate values for the thermal conductances of the heat exchangers, as well as errors in the prediction of the dependence of chiller performance upon key variables such as cooling rate and coolant temperatures. In Section B, these claims were demonstrated for a reciprocating chiller. We will now use experimental measurements, in combination with computer simulation calculations and the fundamental thermodynamic chiller model, to establish these points for an absorption system, including examples of the implications for chiller diagnostics and optimization.

For absorption chillers, internal losses connote some processes that are not present in mechanical chillers. The internal irreversibilities that are unique to absorption chillers stem from: (1) losses in the chemical potentials of the refrigerant and solution as a consequence of finite-rate mass transfer (chemical potential drop); and (2) all losses in regenerative heat exchangers being internal, because their heat exchange involves no thermal communication with the heat reservoirs. In addition, as in mechanical chillers, pressure-drop losses in the heat exchangers may not be negligible.

C2. About regenerative absorption chillers

Now we'll illustrate several points related to absorption chillers with a specific case study. Our analysis is based in part on experimental measurements reported in [Taniguchi *et al* 1996] for an ammonia–water absorption chiller (illustrated schematically in Figure 12.3). A typical absorption chiller can have as many as 3 additional regenerative (internal) heat exchangers at the: (1) generator (GHE), (2) absorber (AHE), and (3) generator–absorber (GAX) interface. The potential for decreasing internal entropy production, and thereby increasing chiller COP, with heat regeneration (internal heat recovery) is considerable because of the typically large temperature differences across the strong and weak solutions. This point is illustrated graphically in Figure 12.4, which is a pressure–temperature or so-called Duehring diagram of the chiller cycle.

The Duehring diagram is a common engineering representation of the thermodynamic state information in the absorption cycle. The Duehring diagram also clearly indicates the potential for heat regeneration in the internal heat exchangers (GAX, AHE and GHE). It merits qualifications, however, because a convention used in the Duehring diagram does not correspond to the actual physical processes involved. Specifically,

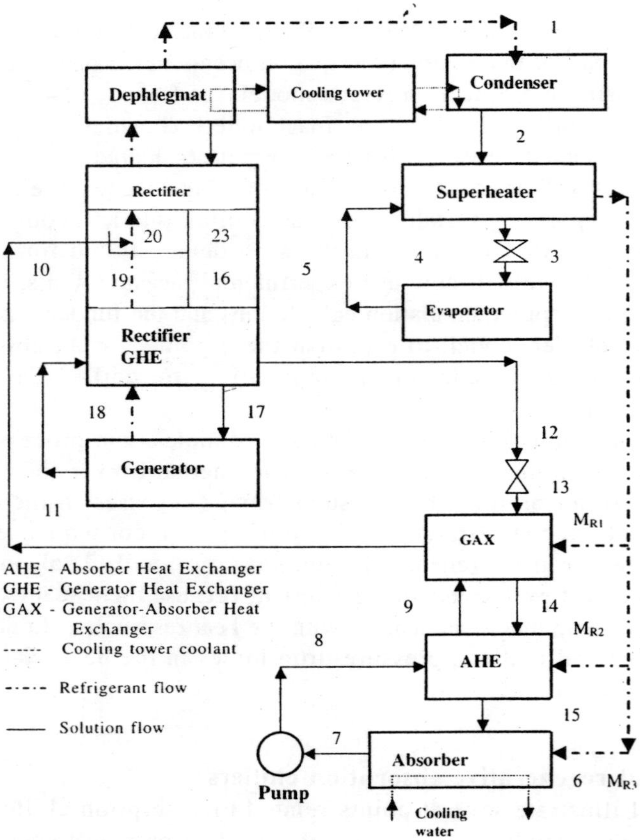

Figure 12.3: Schematic of a typical ammonia-water GAX absorption chiller cycle. State points are numbered in Figure 12.4 for later reference in Table 12.5.

each of the isosteres (paths of constant concentration, labeled in Figure 12.4 with XR, X1 or X2 and solid-head arrows) in reality is not a single thermodynamic path. Rather, each isostere comprises two paths: isobaric heat transfer and throttling (or vice versa depending on where in the cycle it occurs). The isosteres are drawn only as a means of condensing the information on concentration along the paths. Along the actual path, the fluid has the same concentration as that described by the saturation isostere. The sloped (as opposed to vertical or horizontal) broken lines in Figure 12.4 are fictitious and are included to demarcate clearly the regions of regeneration. While the state points cited are correct, the paths could be misleading were one to model the cycle solely from the Duehring diagram.

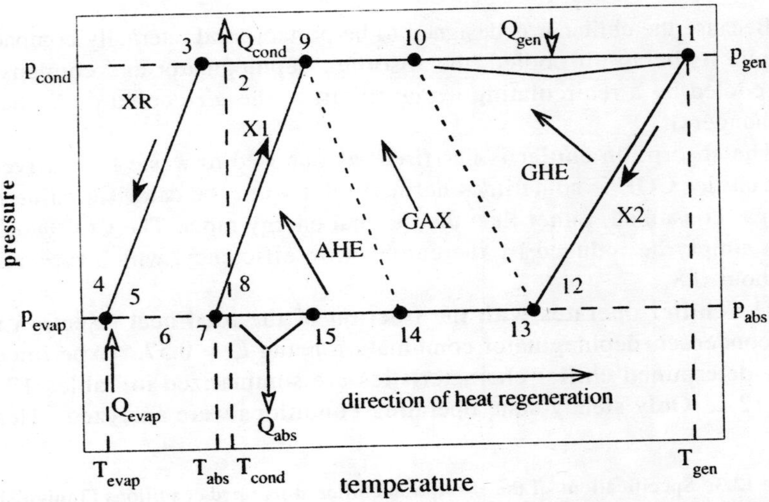

Figure 12.4: Pressure-temperature or so-called Duehring diagram of a typical ammonia-water absorption chiller cycle with regenerative heat exchangers. States are numbered as in Figure 12.3. Open-head arrows indicate the direction of regenerative heat flows. Solid-head arrows indicate the 3 isosteres (paths of constant concentration). White-head arrows indicate heat flows at the heat exchangers.

C3. Experimental details

The air-cooled ammonia–water absorption chiller studied here has a nominal rated cooling rate of 7 kW, at which point the COP is 0.83. A rectifier and dephlegmator are included to prevent water vapor from entering the condenser. Since a small fraction of the heat rejection is then effected at the dephlegmator (the major heat rejection being at the condenser and absorber), it must be accounted for explicitly in the entropy and entropy balances, and hence in the PAT calculations. This is a straightforward exercise that follows the formalism we established in Chapter 4. The dephlegmator (subscript dep) and condenser are cooled in series. In the energy balance equations (4.25) and (4.27), Q_{cond} is replaced by $Q_{cond} + Q_{dep}$. The entropy balance equation (4.26) has an additional term Q_{dep}/T_{dep}. And in the chiller performance equation (4.28) and PAT equations (4.31) and (4.32), T_{cond} should be interpreted as the entropically-weighted average of the condenser and dephlegmator PATs:

$$\frac{1}{T_{cond}} = \frac{\dfrac{Q_{dep}}{\text{PAT}_{dep}} + \dfrac{Q_{cond}}{\text{PAT}_{cond}}}{Q_{dep} + Q_{cond}}. \tag{12.2}$$

Because the chiller was designed to be portable and internally compact, despite its being air cooled, the absorber, dephlegmator and condenser are cooled by a recirculating *water* circuit to the *air*-cooled coils (heat exchangers).

The absorption chiller is gas-fired, as opposed to waste-heat driven. The chiller COP is sometimes defined relative to the calorific value of the gas consumed, rather than the thermal energy input. The COP would then simply be reduced by the combustion efficiency, which typically is about 0.8.

The chiller operates with the fraction of the total heat rejection at the condenser+dephlegmator combination being $\xi = 0.57$. Experimentally-determined chiller characteristics are summarized in Tables 12.3 and 12.4. Only steady-state operating conditions were analyzed. Heat

Table 12.3: Specifications of the absorption chiller at its rated conditions [Taniguchi *et al* 1996].

chiller component (state points in parentheses – see Figures 12.3 and 12.4)	value and units
refrigerant outlet temperature from the evaporator (5)	5°C
refrigerant outlet temperature from the condenser (2)	44°C
refrigerant subcooling temperature after the condenser	25°C
refrigerant outlet temperature from the dephlegmator (1)	76°C
strong solution outlet temperature from the absorber (7)	41°C
weak solution outlet temperature from the generator (13)	194°C
pressure	
generator	18.0 bar
absorber	4.8 bar
pressure drop on the low pressure part of the cycle	0.3 bar
pressure drop on the high pressure part of the cycle	0.6 bar
thermal conductances for latent heat transfer	
regenerative generator–absorber heat exchanger (GAX)	257 W K^{-1}
absorber heat exchanger (AHE)	219 W K^{-1}
dephlegmator	30 W K^{-1}
superheater	47 W K^{-1}

Table 12.3 continued

thermal conductances for sensible heat transfer	
absorber	398 W K^{-1}
condenser	1434 W K^{-1}
evaporator	984 W K^{-1}
generator	255 W K^{-1}
solution flow rates	
strong solution (ammonia–water)	48.5 kg h^{-1}
refrigerant (99% ammonia)	22.8 kg h^{-1}
cooling rate	6.89 kW

Table 12.4: Thermodynamic states for the absorption chiller cycle

State	T (°C)	p (bar)	ammonia (refrigerant) mass fraction, X	specific enthalpy, h (kJ kg^{-1})	specific entropy, s (kJ kg^{-1} K^{-1})	m (kg s^{-1})
1	76.24	17.95	0.999	1394	4.429	0.00606
2	44.00	17.35	0.999	211	0.720	0.00606
3	25.00	17.35	0.999	121	0.426	0.00606
4	5.00	5.02	0.999	121	0.453	0.00606
5	5.00	5.02	0.999	1268	4.571	0.00606
6	38.36	5.02	0.999	1359	4.922	0.00606
7	41.00	4.72	0.476	−60	0.420	0.01347
8	41.00	18.75	0.476	−60	0.420	0.01347
9	90.66	18.75	0.476	171	1.098	0.01347
10	106.99	18.55	0.398	253	1.312	0.01347
11	194.00	17.95	0.047	801	2.325	0.00742
12	134.14	17.95	0.047	535	1.720	0.00742
13	134.14	4.72	0.047	535	1.720	0.00742
14	95.66	4.72	0.197	275	1.251	0.00880

Table 12.4 continued

15	72.19	4.72	0.306	116	0.900	0.01096
16	106.99	17.95	0.398	253	1.312	0.01202
17	145.89	17.95	0.226	489	1.818	0.01096
18	169.95	17.95	0.600	2064	5.816	0.00354
19	106.99	17.95	0.963	1512	4.754	0.00460
20	106.99	17.95	0.963	1512	4.754	0.00645
21	100.48	17.95	0.974	1483	4.678	0.00655
22	76.24	17.95	0.571	115	0.922	0.00049
23	106.99	17.95	0.398	253	1.312	0.00039

Summary of predicted heat transfer rates and performance variables

Q_{evap} = 6.89 kW	Q_{cond} = 7.17 kW	Q_{abs} = 6.47 kW	Q_{gen} = 8.28 kW
Q_{dep} = 1.21 kW	Q_{AHE} = 3.11 kW	Q_{GAX} = 3.44 kW	Q_{GHE} = 1.93 kW
COP = 0.841			
ΔS_{int} = 2.86 W K^{-1}			
ξ = 0.57			

is supplied from the generator at 200°C. The design range is T_{cond}^{out} = 37–44°C and T_{evap}^{out} = 5–7°C.

C4. Calculation of the PATs and internal entropy production

In those earlier studies of absorption chillers in which the significance of the PAT was recognized but a proper procedure for quantifying the PAT was not developed, the PAT was approximated as the outlet temperature of the coolant or of the refrigerant for a given process. These approximations excluded internal entropy production in the heat exchangers. As will be demonstrated below, internal dissipation comprises about half of the total (external plus internal) entropy production in the chiller. Furthermore, the omission of internal dissipation in the heat exchangers constitutes neglecting the lion's share of internal losses. Unlike the treatment in Section B for a mechanical chiller, the proper PAT in absorption systems must account not only for the contribution of pressure drop (fluid friction), but also for chemical potential drops and for the irreversibilities of internal heat recovery.

The PATs of the absorber and generator heat exchangers, as described by Equations (4.4)–(4.12), can be calculated as follows. One integrates along the bubble and dew lines of the binary mixture phase diagram, and takes into account the initial superheating of the refrigerant stream and the strong solution stream in the absorber and generator, respectively. Irreversibilities deriving from finite-rate mass transfer are comparatively small to the extent that the actual phase change deviates from that described by the binary mixture phase change diagram. One then treats the amount of water vapor entering the condenser as negligible, due to the installation of a rectifier and dephlegmator.

The PATs of the condenser and evaporator heat exchangers are calculated by integrating along the isobars of pure ammonia. The general procedure for calculating PATs and ΔS_{int} for absorption chillers is illustrated in Tutorial 4.2 for the specific instance of a LiBr–water pair. With a simulation procedure, the method can be adapted to the volatile ammonia–water solution.

With the expression for heat transfer rate Q_i at heat exchanger i, the thermal conductance $(mCE)_i$ is calculated from

$$Q_i = \sum_{\beta=1}^{j}\left(m_\beta h_\beta\right)_{\text{in}} - \sum_{\beta=1}^{k}\left(m_\beta h_\beta\right)_{\text{out}} = (mCE)_i\left(T_i^{in} - \text{PAT}_i\right) \tag{12.3}$$

where β denotes each of the autonomous phases, and there are a total of j inlet streams and k outlet streams.

Next we express the internal entropy production $\Delta S_{\text{int}}^{(i)}$, in two equivalent forms, as noted in Equations (4.4) and (4.6). In the general form, we have

$$\Delta S_{\text{int}}^{(i)} = \sum_{\beta=1}^{j}\left(m_\beta s_\beta\right)_{\text{in}} - \sum_{\beta=1}^{k}\left(m_\beta s_\beta\right)_{\text{out}} + \frac{Q_i}{\text{PAT}_i}. \tag{12.4}$$

For the 3 regenerative heat exchangers, the contribution of the Q_i term in Equation (12.4) is taken as zero since, from the viewpoint of internal losses, internal heat recovery can be viewed as adiabatic. Accounting explicitly for the pressure-drop and chemical-potential-drop losses, we also have

$$\Delta S_{\text{int}}^{(i)} = -\sum_{\text{all streams}\,\beta} \int_{\text{in}}^{\text{out}} \left[\frac{m_\beta v_\beta \, dp}{T_\beta} + \frac{\mu_\beta \, dm_\beta}{T_\beta} \right] \qquad (12.5)$$

with $\Delta S_{\text{int}}^{(i)}$ always being positive.

C5. Computer simulation formulation and validation

Our strategy was to develop a computer simulation that, for the purposes of this study, could substitute for the chiller itself, and then to perform sensitivity and optimization studies. First, however, we need to establish the validity of the simulation by showing that it can accurately predict each principal aspect of experimentally measured chiller performance.

We assembled a computer simulation of the chiller from component-by-component routines that are summarized in Table 12.5. The routines are approximate rather than rigorous. Each individual component may be modeled, but it is modeled effectively as a blackbox of known average thermodynamic parameters. For example, heat transfer and mass transfer resistances are absorbed into the calculation of the PATs. Heat exchangers are characterized by a single UA value, rather than properly treating their UA values as varying along the length of the heat exchanger. The equations specific to the determination of the enthalpy and entropy of the ammonia–water solution follow the established procedures of [Ziegler & Trepp 1984; Herold *et al* 1996].

The comparisons for validating the accuracy of the simulation are presented in Table 12.6. In Table 12.6, the typical experimental uncertainty for determining the principal heat flows (the Q's) is 5–7%. The relatively large discrepancy for the dephlegmator is caused by the uncertainty in its construction and therefore in its overall heat transfer coefficient. Nonetheless, the error in the total energy output (*i.e.*, the sum of heat flows from the absorber, condenser and dephlegmator) is within experimental uncertainty.

C6. Quantitative results for internal dissipation and the implications

Having found reasonable agreement between the simulation predictions and experimental measurements, we now use the simulation to explore chiller behavior in a manner that would be extremely difficult in the laboratory. Table 12.7 is a succinct summary of the relative contributions of each chiller element to internal losses, as well as the

Table 12.5: Governing equations for heat transfer rate, internal entropy generation rate and mass balance at each chiller component. Subscript numbers refer to the state points drawn in Figure 12.3.

component (see Figure 12.3)	rate of heat transfer Q_i	rate of internal entropy generation $\Delta S^{(i)}_{int}$	mass balance equations
condenser (path 1–2)	$Q_{cond} = m_1(h_1 - h_2)$	$m_1(s_2 - s_1) + \frac{Q_{cond}}{PAT_{cond}}$	$m_1 = m_2$
evaporator (path 4–5)	$Q_{evap} = m_5(h_5 - h_4)$	$m_5(s_5 - s_4) - \frac{Q_{evap}}{PAT_{evap}}$	$m_4 = m_5$
absorber (paths 15–7, 6–7)	$Q_{abs} = m_{15}h_{15} - m_{R3}h_6 - m_7h_7$	$m_7s_7 - m_{15}s_{15} - m_{R3}s_6 + \frac{Q_{abs}}{PAT_{abs}}$	$\frac{m_{R3}}{m_{15}} = \frac{X_7 - X_{15}}{X_1 - X_7}$, $m_7 = m_{15} + m_{R3}$
generator (paths 17–11, 17–18)	$Q_{gen} = m_{11}h_{11} + m_{18}h_{18} - m_{17}h_{17}$	$m_{11}s_{11} - m_{17}s_{17} + m_{18}s_{18} - \frac{Q_{gen}}{PAT_{gen}}$	$\frac{m_{18}(Y_{18} - X_{17})}{17} = -m_{16} - m_{19} = \frac{m_{16}X_{16} - m_{19}Y_{19}}{X_{17}}$, $m_{17} = m_{16} + m_{18} - m_{19}$
dephlegmator* (paths 21–1, 21–22)	$Q_{dep} = m_{20}h_{10.V} - m_1h_1 - m_{23}h_{10.L}$	$m_1s_1 + m_{22}s_{22} - m_{21}s_{21} + \frac{Q_{dep}}{PAT_{dep}}$	$\frac{m_{22}\eta_d}{m_1} = \frac{X_1 - Y_{21}}{Y_{21} - X_{22}}$, $m_{21} = m_{22} + m_1$
rectifier (paths 18–21, 22–17)		$\sum_{out} m_is_i - \sum_{in} m_is_i$	$\frac{m_{23}}{m_{19}} = \frac{X_1 - Y_{10}}{Y_{10} - X_{10}}$, $m_{20} = m_{23} + m_2$, $m_{19} = m_{20} + m_{10,V}$, $m_{16} = m_{23} + m_{10,L}$
regenerative generator heat exchanger GHE (paths 11–12, 19–18, 16–17)	$Q_{GHE} = m_{11}(h_{11} - h_{12})$	$\sum_{out} m_is_i - \sum_{in} m_is_i$	$X_{10} = X_{max}$
regenerative absorber heat exchanger AHE (paths 8–9, 14–15), 6 –15)	$Q_{AHE} = m_9(h_{10} - h_9)$	$\sum_{out} m_is_i - \sum_{in} m_is_i$	$\frac{m_{R2}}{m_{14}} = \frac{X_{15} - X_{14}}{X_1 - X_7}$, $m_{15} = m_{14} = m_{R2}$

Table 12.5: continued

regenerative generator–absorber heat exchanger GAX (paths 9–10, 13–14, 6–14)	$Q_{GAX} = m_{13}h_{13} + m_{R1}h_6 - m_{14}h_{14}$	$\sum_{out} m_i s_i - \sum_{in} m_i s_i$	$\frac{m_{19}}{m_1} = \frac{X_1 - X_{min}}{X_{max} - X_{min}}$, $\frac{m_{13}}{m_1} = \frac{X_1 - X_{max}}{X_{max} - X_{min}}$, $m_{R1}(X_{max} - X_{min}) = \frac{X_{14} - X_{min}}{X_1 - X_{16}}$, $\frac{m_{14} - m_{R1}}{m_1} = \frac{X_1 - X_{max}}{X_{max} - X_{min}}$
superheater (paths 2–3, 5–6)	$Q_{super} = m_2(h_2 - h_3) = m_5(h_6 - h_5)$	$m_2(s_3 + s_6 - s_2 - s_5)$	$m_2 = m_3$, $m_5 = m_6$
valve (between the superheater and the evaporator) (path 3–4)	$Q_{valve} = m_3h_3 = m_4h_4$ (due to throttling)	$m_3(s_4 - s_3)$	$m_3 = m_4$

X_j's = refrigerant concentration in the liquid. Y_j's = refrigerant concentration in the vapor.
V = vapor. L = liquid.
R1 = refrigerant mass flow rate to GAX.
R2 = refrigerant mass flow rate to AHE.
R3 = refrigerant mass flow rate to absorber.
min = minimum concentration of ammonia–water solution.
max = maximum concentration of ammonia–water solution.
* To correct for the fact that the outlet vapor concentration in the dephlegmator can be less than the value calculated under ideal conditions, we introduce a safety factor η_d (taken here as 0.8) in the mass balance equation to ensure that in reality the condensate flow rate will indeed yield the prescribed concentration. This is common engineering practice because chiller performance is crucially sensitive to the concentration of refrigerant.

magnitudes of external and internal entropy production. In light of heat exchanger internal entropy production having been excluded from earlier thermodynamic models of absorption chillers, the following points are of particular note:

(1) Internal entropy production constitutes more than half the total entropy production.

(2) Internal losses are overwhelmingly dominated by heat exchanger dissipation.

(3) Not surprisingly, the heat exchangers that span the largest temperature and chemical potential differences contribute the largest entropy generation.

Analyses that ignore heat exchanger internal losses use an incorrect PAT and hence can provide misleading predictions for the sensitivity of chiller COP to the principal control variables. To illustrate this point for the absorption chiller, we list in Table 12.8 the internal entropy production rate, and the external heat exchanger thermal conductances,

Table 12.6: Comparison of predictions of the computer code against experimental measurements for the absorption chiller. Inlet coolant water temperature = 31.9°C.

property	predicted	measured
condenser pressure	17.95 bar	18.0 bar
evaporator pressure	4.72 bar	4.8 bar
refrigerant outlet temperature at the absorber	41°C	40°C
refrigerant outlet temperature at the generator	194°C	193°C
refrigerant concentration	99.9%	99.9%
strong solution concentration	47.56%	47.4%
weak solution concentration	4.71%	4.7%
Q_{gen}	8.28 kW	8.37 kW
heat rejection at the dephlegmator, Q_{dep}	1.21 kW	2.03 kW
Q_{evap}	6.96 kW	6.89 kW
Q_{cond}	7.17 kW	6.85 kW
Q_{abs}	6.47 kW	6.12 kW
heat transfer rate across the regenerative generator heat exchanger Q_{GAX}	3.44 kW	3.40 kW
heat transfer rate across the regenerative absorber heat exchanger Q_{AHE}	3.11 kW	3.10 kW
COP	0.841	0.823

deduced with the correct PAT, and with a PAT based upon the omission of heat exchanger internal dissipation, at the fixed coolant temperatures at rated conditions.

We then determine the operating conditions (specifically, the ξ value) at which COP is maximized at the particular cooling rate (Q_{evap} = 6.89 kW) and coolant temperatures. The proximity of the actual chiller COP of 0.823 to the optimum of 0.844 attests in part to the empirical wisdom embodied in chiller development.

When the actual chiller performance data are used in the analytic model, neglecting heat exchanger internal losses leads to artificially high

Table 12.7: Contribution of individual chiller components to internal entropy production. *External* chiller entropy production rate = 2.77 W K^{-1} (which includes 0.31 W K^{-1} from heat leaks at the generator and evaporator). *Internal* entropy production rate = 2.86 W K^{-1} (51% of the total). The figures below refer to the fraction of the *internal* entropy production.

component	**relative contribution to the *internal* entropy production**
generator and regenerative generator heat exchanger*	0.376
regenerative generator–absorber heat exchanger	0.235
regenerative absorber heat exchanger	0.168
valve between superheater and evaporator	0.057
absorber	0.053
superheater	0.049
condenser	0.046
evaporator	0.010
dephlegmator	0.006

*The *regenerative* generator heat exchanger GHE is built within the generator. Although the generator and the GHE can be modeled separately in the computer code, experimental measurements can reveal only the properties of the combination of the generator and GHE.

internal entropy production. This in turn results in deducing an incorrectly high heat exchanger thermal conductance, along with an unduly high COP. In diagnostic studies, this could result in errors in evaluating the change with time of heat exchanger quality or of internal losses. There would also be errors in determining the best values of cooling rate and of ξ.

Table 12.9 offers a comparison between: (1) COP and ξ values at assorted cooling rates for the actual chiller; and (2) the maximum realizable COP for each of these cooling rates, and the associated ξ value (*i.e.*, the nominal optimum). The fact that absorption chiller technology has empirically evolved to what can now be discerned as near-optimal operating conditions is striking. For the practical range of cooling rates from full load to about 60% part load, the actual COPs are within 1% of the maximum attainable values. Furthermore, the COP and ξ values at full load differ by only a few percent from the part-load values at which COP reaches its global optimum.

Table 12.8: Characteristics of the optimized chiller (*i.e.*, to achieve maximum COP), determined with the correct PAT (*i.e.*, accounting for heat exchanger internal dissipation), and with an uncorrected PAT equal to the outlet temperature of the refrigerant or solution.

property	with the correct PAT	with an uncorrected PAT
ΔS_{int}	2.86 W K^{-1}	5.09 W K^{-1}
$(mCE)_{evap}$	984 W K^{-1}	984 W K^{-1}
$(mCE)_{gen}$	255 W K^{-1}	1315 W K^{-1}
$(mCE)_{abs}$	398 W K^{-1}	2156 W K^{-1}
$(mCE)_{cond}$	1434 W K^{-1}	1675 W K^{-1}
optimal COP	0.844	0.911
Q_{evap}	7.03 kW	6.50 kW
ξ	0.59	0.77

C7. Qualifications

Clearly, the blackbox approach to chiller analysis has its limitations when precise predictions of key chiller variables are essential, especially given the complexities of absorption machines. Although the computer routine described in this section goes component-by-component, the approach remains basically a blackbox method. The predictive and diagnostic accuracy we've examined in this chapter may be adequate for many needs of cooling engineers, but still leaves non-negligible room for improvement. In addition, the more fastidious reader may prefer a *completely* self-consistent distributed modeling scheme.

One important limitation assumption in our analysis is that we *assume* a thermodynamic state for the refrigerant vapor at the outlet of the generator. Often, the value of the heat exchanger's effectiveness E or overall thermal conductance UA is estimated based on this assumed

Table 12.9: Actual chiller COP and ξ values at selected cooling rates compared against the maximum attainable COP, and the associated ξ value, at the same cooling rates.

Q_{evap}(kW)	COP		ξ	
	actual operation	max. value	actual operation	at max. COP
6.890	0.823	0.844	0.570	0.590
6.234	0.836	0.856	0.577	0.601
5.578	0.848	0.864	0.586	0.613
4.921	0.856	0.869	0.594	0.619
4.265	0.861	0.873	0.608	0.633

outlet state. In reality, however, mass transfers proceed non-isothermally throughout the generator and absorber. Furthermore, when outlet states are pre-defined, the effects of mass-transfer resistance are automatically excluded. So although the methods and results we've developed for absorption chillers in this chapter are relatively simple to implement, and produce acceptably accurate predictions of chiller behavior for many uses, they are not rigorous.

Quite recently, a rigorous detailed model for absorption chillers has been developed [Chua 1999], such that local thermodynamic balance for all heat and mass transfers, within all chiller components, is respected. This represents a step beyond simply insuring an overall thermodynamic balance between the inlet and outlet states of the chiller components (most notably the generator and absorber, where the major mass-transfer bottlenecks occur).

Chapter 13

TEMPERATURE–ENTROPY DIAGRAMS FOR REPRESENTING REAL IRREVERSIBLE CHILLERS

"For every complex problem, there is a solution that is simple, neat and wrong."
- H.L. Mencken

A. BACKGROUND

A thermodynamic diagram for a chiller cycle is usually plotted as pressure against volume, or temperature against entropy, for the refrigerant, because the areas under these curves are readily identifiable with heat flows and work. Particularly convenient and instructive is the T–S diagram for *ideal* chiller cycles, where all heat transfers are isothermal and all connecting branches (*e.g.*, compression and expansion) are isentropic (Figure 2.2). It is then easy to visualize each of the principal elements in the energy and entropy balance, not to mention the facility with which they can be calculated as the areas of simple rectangles.

A classic pedagogical example is the reversible cycle (subscript "rev"), drawn in Figure 13.1. Figure 13.1 is embellished to include another idealized instructive example: the endoreversible cycle (subscript "endo"), in which the external irreversibilities of finite-rate heat transfer are incorporated (but internal dissipation is neglected). T' and T denote reservoir and refrigerant temperatures, respectively.

Figure 13.1 is drawn such that the cooling capacity is fixed for *both* the reversible and the endoreversible cycles, as would be common in standard engineering practice. Differences between the cycles are then manifested in the differences in power input and heat rejection.

Referring to Figure 13.1, we can immediately identify the following key variables:

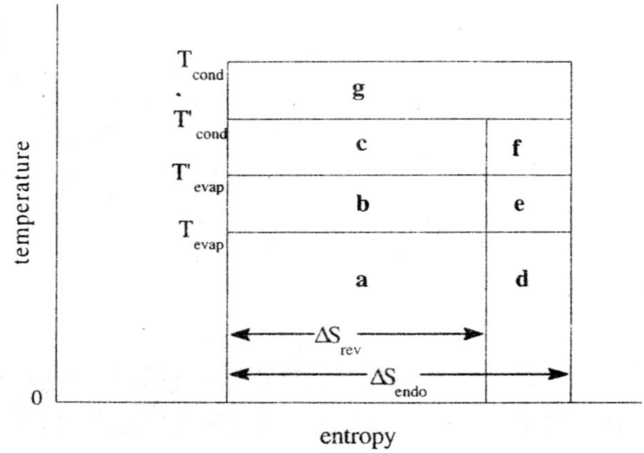

Figure 13.1: *T–S* diagram for an idealized chiller. The cycle comprises isotherms connecting isentropes. The reversible cycle operates between temperatures T'_{cond} for heat rejection and T'_{evap} at the cooling load, and across an entropy difference ΔS_{rev}. The corresponding endoreversible cycle functions between T_{cond} and T_{evap}, and across an entropy difference ΔS_{endo}.

variable	reversible cycle	endoreversible cycle
heat rejection Q_{cond}	sum of rectangles **a + b + c**	sum of rectangles **a + b + c + d + e + f + g**
cooling capacity Q_{evap}	sum of rectangles **a + b**	sum of rectangles **a + d** $\left(\text{with } Q_{evap}^{rev} = Q_{evap}^{endo}\right)$
work input $P_{in}=Q_{cond}-Q_{evap}$	rectangle **c**	sum of rectangles **b + c + e + f + g**
Coefficient of Performance $COP=Q_{evap}/P_{in}$	$\frac{\mathbf{a+b}}{\mathbf{c}}$	$\frac{\mathbf{a+d}}{\mathbf{b+c+e+f+g}}$
total external losses	none	sum of rectanges **d + e + f + g**

A qualification is in order regarding the consistency of units for the variables considered here. In earlier chapters, Q_{evap}, P_{in}, Q_{cond}, ΔS values, etc. have been expressed as cycle-average *rates*. For example, Q_{evap}, Q_{cond} and P_{in} were expressed in kW, ΔS in kW K^{-1}, etc. In this chapter, in order to relate to classic physics and engineering *T–S* diagram representations, we express energy flows as specific energies in units of kJ kg^{-1}, and entropy flows as specific entropy flows in units of kJ kg^{-1} K^{-1}. Specific refers to per unit mass of refrigerant. By multiplying these specific quantities by the refrigerant mass flow rate (in units of kg s^{-1}), one obtains the energy or entropy rates used in previous chapters.

From Figure 13.1, we can also picture the confirmation of the Second Law:

$$\frac{Q_{\text{cond}}}{T_{\text{cond}}} - \frac{Q_{\text{evap}}}{T_{\text{evap}}} \geq 0. \tag{13.1}$$

Each of the two terms on the left-hand side of Equation (13.1) is equal to the abscissa range ΔS_{endo}, and the equality (as opposed to the inequality) in Equation (13.1) is realized because the model does not account for internal dissipation.

As convenient and instructive as this exercise may be, it does not relate to the actualities of genuinely irreversible chillers. As we've seen in earlier chapters, real chillers certainly incur external irreversibilities, but they are dominated by internal dissipation. Furthermore, their heat transfers are not isothermal; neither are their compression and expansion branches isentropic. Is it possible, then, to represent actual irreversible cycles in the calculationally and visually convenient form of the rectangular T–S diagram of Figure 13.1 without compromising any of the physics of the problem? Can one still emerge with an accurate quantitative accounting of all energy and entropy flows while retaining the simple physical picture of Figure 13.1?

The purpose of this chapter is to demonstrate that the answers to these queries are affirmative, and to illustrate the thermodynamic diagram method with real experimental data for actual chillers [Gordon *et al* 1999]. The key to such seemingly simple T–S diagrams is the PAT for the refrigerant – the reference temperature for translating irreversibility (entropy production) into lost work (refer back to Chapter 4). The thermodynamic processes that occur within a chiller may not be isothermal; but one can cast them as effectively isothermal paths with a refrigerant Process Average Temperature (PAT) that yields the correct energy and entropy balances.

For ease of conceptual clarity and flow of logic, we will first address mechanical chillers. Then we will progress to the case of absorption machines.

B. PAT AND THE PERFORMANCE CHARACTERISTIC FOR MECHANICAL CHILLERS

To relate to real irreversible chiller cycles, we need to augment the endoreversible chiller model noted above in two essential ways: (1) the actual non-isothermal heat transfers must be expressed in terms of effectively isothermal heat transfers; and (2) internal losses must be incorporated. The PAT relates to both points. Formally, it can be expressed for each individual process as (recall Equations (4.10)–(4.11))

$$\mathrm{PAT} = \frac{\int \mathrm{d}H}{\int \mathrm{d}S - \Delta S_{\mathrm{int}}} = \frac{\int \mathrm{d}H}{\int \frac{\mathrm{d}H}{T}}. \tag{13.2}$$

Our objective is to relate to cyclic chiller performance within a relatively simple *T–S* diagram (or, more properly, a PAT–*S* diagram) in which thermodynamic performance can be expressed in terms of effective isotherms connected by effective isentropes. Hence the integrals in Equation (13.2) are performed over all processes within the chiller and lumped into two PAT values: $\mathrm{PAT}_{\mathrm{high}}$ and $\mathrm{PAT}_{\mathrm{low}}$.

Now recall the characteristic chiller performance formula derived in Chapter 4:

$$\begin{aligned}\frac{1}{\mathrm{COP}} &= \left[-1+\frac{\mathrm{PAT}_{\mathrm{high}}}{\mathrm{PAT}_{\mathrm{low}}}\right]+\frac{\mathrm{PAT}_{\mathrm{high}}\,\Delta S_{\mathrm{int}}}{Q_{\mathrm{evap}}}+\frac{\mathrm{PAT}_{\mathrm{high}}\,\Delta S_{\mathrm{leak}}}{Q_{\mathrm{evap}}}\\ &= \frac{1}{\mathrm{COP}_{\mathrm{rev}}}+\frac{1}{\mathrm{COP}_{\mathrm{int}}}+\frac{1}{\mathrm{COP}_{\mathrm{leak}}}\end{aligned} \tag{13.3}$$

with

$$\mathrm{PAT}_{\mathrm{high}} = T_{\mathrm{cond}} \qquad \mathrm{PAT}_{\mathrm{low}} = T_{\mathrm{evap}}$$

and

$$\Delta S_{\mathrm{leak}} = \frac{Q_{\mathrm{comp}}^{\mathrm{leak}}}{T_{\mathrm{cond}}} + Q_{\mathrm{evap}}^{\mathrm{leak}}\left[\frac{1}{T_{\mathrm{evap}}} - \frac{1}{T_{\mathrm{cond}}}\right]. \tag{13.4}$$

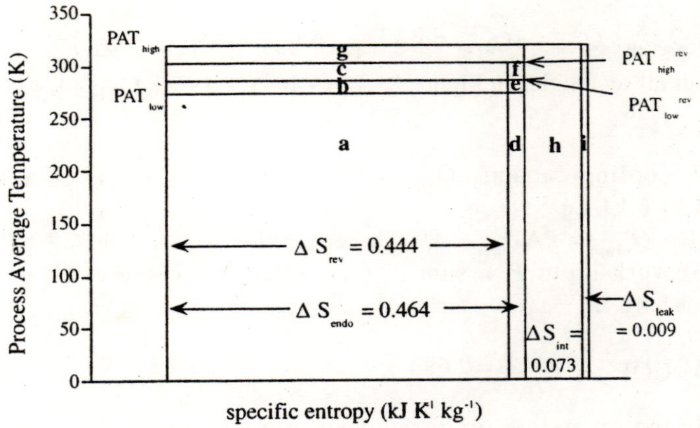

Figure 13.2: PAT–entropy diagram for a real irreversible reciprocating chiller, based on experimental measurements.

We can now represent the key quantitative aspects of the chiller cycle in a stacked-rectangle PAT–*S* diagram, as drawn in Figure 13.2 – a representation we'll now explore.

C. PAT–ENTROPY DIAGRAM FOR MECHANICAL CHILLERS

The type of extensive and accurate experimental measurements needed for our illustrative calculations were reported for a water-cooled vapor-compression reciprocating chiller in Chapter 6. The chiller rated cooling rate is 10.66 kW at which point its COP is 2.68. The rated condition has T^{in}_{evap}= 12.4°C and T^{in}_{cond}= 29.4°C. The refrigerant flow rate is 0.084 kg s^{-1}. Details of the PAT calculations were presented in Tutorial 4.1. The internal, external and heat leak losses were evaluated in Chapter 6.

The PAT–entropy diagram in Figure 13.2 highlights the critical role of internal losses. For this reciprocating chiller, $PAT_{high} = T_{cond}$ = 319.22 K; and $PAT_{low} = T_{evap}$ = 273.29 K. The principal energy and entropy balances, and their impact on chiller performance, can be seen in terms of the relative areas of the assorted rectangles in Figure 13.2, as follows.

A) For the overall energy balance:

1) net heat rejection from the chiller =

$$Q_{cond} + Q_{cond}^{leak} + Q_{comp}^{leak} - Q_{evap}^{leak} = PAT_{high}(\Delta S_{endo} + \Delta S_{int} + \Delta S_{leak}) =$$

sum over all rectangles in Figure 13.2 $(\mathbf{a}+\mathbf{b}+\mathbf{c}+\mathbf{d}+\mathbf{e}+\mathbf{f}+\mathbf{g}+\mathbf{h}+\mathbf{i}) =$

$= 174.6$ kJ kg^{-1}

2) cooling capacity $Q_{evap} = PAT_{low}\ \Delta S_{endo}$ = rectangles **a** + **d** = 126.7 kJ kg^{-1}

also $Q_{evap} = PAT_{low}^{rev}\ \ \Delta S_{rev}$ = rectangles **a** + **b** = 126.7 kJ kg^{-1}.

3) work input P_{in}= sum of rectangles **b** + **c** + **e** + **f** + **g** + **h** + **i** = 47.3 kJ kg^{-1}.

4) COP $= \dfrac{Q_{evap}}{P_{in}} = 2.68.$

B) In the reversible limit, the corresponding reservoir temperatures are denoted by PAT_{high}^{rev} and PAT_{low}^{rev}, which are T_{cond}^{in} and T_{evap}^{in}, respectively. The additional performance variables are:

1) minimum work input P_{in}^{min}= rectangle **c** = 7.6 kJ kg^{-1}.

2) $COP_{rev} = \dfrac{Q_{evap}}{P_{in}^{min}} = 16.8.$

C) Losses stemming from irreversibilities:

1) total external losses = rectangles **d** + **e** + **f** + **g** = 13.7 kJ kg^{-1}.

2) internal losses = rectangle **h** = 23.3 kJ kg^{-1}.

3) heat leak losses = rectangle **i** = 2.8 kJ kg^{-1}.

4) total losses = 39.8 kJ kg^{-1}, of which 54% are solely internal losses.

The endoreversible chiller model predicts

$$COP_{endo} = \frac{Q_{evap}}{(PAT_{high} - PAT_{low})\Delta S_{endo}} = 5.95$$

which is a factor of 2.2 higher than the actual COP.

In addition, it is graphically clear from Figure 13.2 that the Second Law is respected:

$$\frac{\text{net heat rejection}}{PAT_{high}} - \frac{Q_{evap}}{PAT_{low}} = \Delta S_{int} + \Delta S_{leak} \geq 0 \qquad (13.5)$$

with the inequality (rather than the equality) applying in Equation (13.5), as a consequence of internal and heat leak losses.

D. PAT AND THERMODYNAMIC DIAGRAMS FOR ABSORPTION CHILLERS

In Chapter 4, we derived the characteristic performance curve for absorption chillers, which is precisely Equation (13.3), but with PAT_{high} and PAT_{low} expressed in terms of the refrigerant PATs at the four reservoirs, T_{gen}, T_{cond}, T_{abs} and T_{evap}, and in terms of the fraction ξ of the total heat rejection effected at the condenser:

$$\frac{1}{PAT_{high}} = \frac{1}{T_{abs}} - \frac{1}{T_{gen}} - \xi\left[\frac{1}{T_{abs}} - \frac{1}{T_{cond}}\right] \tag{13.6}$$

$$\frac{1}{PAT_{low}} = \frac{1}{T_{evap}} - \frac{1}{T_{gen}}. \tag{13.7}$$

In addition, the expression for ΔS_{leak} must be modified from Equation (13.5) to

$$\Delta S_{leak} = Q_{evap}^{leak}\left[\frac{1}{T_{evap}} - \frac{1-\xi}{T_{abs}} - \frac{\xi}{T_{cond}}\right] - Q_{gen}^{leak}\left[\frac{1}{T_{gen}} - \frac{1-\xi}{T_{abs}} - \frac{\xi}{T_{cond}}\right]. \tag{13.8}$$

The corresponding values for PAT_{high}^{rev} and PAT_{low}^{rev} are obtained by using the individual component *coolant inlet* temperatures, instead of the corresponding process–average refrigerant temperatures, in Equations (13.6) and (13.7).

Experimental measurements the extent and precision of which are adequate for our evaluations were reviewed in Chapters 9 and 12 for a commercial air-cooled ammonia–water absorption chiller with a nominal rated cooling rate of 7.0 kW, and heat rejection divided between the condenser and absorber such that $\xi = 0.57$. At the rated conditions, COP = 0.84. Heat is supplied from the generator at 200°C. The design range of coolant outlet temperatures is 37–44°C at the condenser, and 5–7°C at the evaporator. The evaporator refrigerant flow rate is 0.00637 kg s^{-1}.

The methods for estimating PATs and internal dissipation in absorption chillers were covered in Tutorials 2.3 and 4.2 for the specific case of a LiBr–water system. Applying those methods to the ammonia–water absorption chiller considered here, we can perform the analogous calculations. Before proceeding to the results for the ammonia–water unit, and in the spirit of completing the illustrative examples for the LiBr–water chiller, we take a moment to offer a tutorial that shows the step-by-step procedure for developing the PAT–entropy diagram for the LiBr–water example.

Tutorial 13.1

From the results developed in Tutorials 2.3 and 4.2 for the 3068 kW single-stage LiBr–water absorption chiller, calculate the values of:

(1) the effective hot end and cold end PATs, PAT_{high} and PAT_{low}; and

(2) the endoreversible and internal dissipation contributions to the entropy balance, ΔS_{endo} and ΔS_{int}.

Solution: In addition to quantities that we have already estimated explicitly in Tutorials 2.3 and 4.2, we need the fraction ξ of total heat rejection effected at the condenser:

$$\xi = \frac{Q_{cond}}{Q_{cond} + Q_{abs}} = \frac{3298}{3298 + 4114} = 0.4450.$$

Now

$$\frac{1}{PAT_{high}} = \frac{1}{T_{abs}} - \frac{1}{T_{gen}} - \xi\left[\frac{1}{T_{abs}} - \frac{1}{T_{cond}}\right]$$

$$= \frac{1}{323.43} - \frac{1}{370.20} - 0.4450\left[\frac{1}{323.43} - \frac{1}{319.12}\right] = \frac{1}{2443.8\text{K}}$$

$$\frac{1}{PAT_{low}} = \frac{1}{T_{evap}} - \frac{1}{T_{gen}} = \frac{1}{278.15} - \frac{1}{370.2} = \frac{1}{1118.6\text{ K}}.$$

We calculate ΔS_{endo} from its relation to the given cooling rate:

$$\Delta S_{endo} = \frac{Q_{evap}}{PAT_{low}} = \frac{30680}{1118.6} = 2.743\ \text{kW K}^{-1}.$$

ΔS_{int} is calculated from the relation among the total heat rejection, PAT_{high} and the three principal entropy contributions:

$$Q_{cond} + Q_{abs} = PAT_{high}\,(\Delta S_{endo} + \Delta S_{int} + \Delta S_{leak})$$

where ΔS_{leak} is treated as negligibly small. Hence

$$\Delta S_{int} = \frac{Q_{cond} + Q_{abs}}{PAT_{high}} - \Delta S_{endo} = \frac{3298 + 4114}{2443.8} - 2.743 = 0.2900 \text{ kW K}^{-1}.$$

In Tutorial 4.2, we arrived at an estimate of 0.3007 kW K^{-1} for ΔS_{int}. The difference of 3.7% can be attributed to numerical round-off error where, especially in Tutorial 4.2, one should note that a ΔS_{int} of order 0.3 is obtained from sums and differences of figures of order 10.

The performance of the nominal 7.0 kW ammonia–water absorption chiller can also be summarized in a simple PAT–*S* diagram, as plotted in Figure 13.3. The PAT values turn out to be:

PAT_{high} = 1209.51 K PAT^{rev}_{high} = 858.62 K

PAT_{low} = 748.61 K PAT^{rev}_{high} = 717.65 K.

Qualitatively, almost every aspect of Figure 13.3 is the same as for the reciprocating chiller represented in the PAT–*S* diagram of Figure 13.2. The corresponding summary of the principal energy and entropy balances,

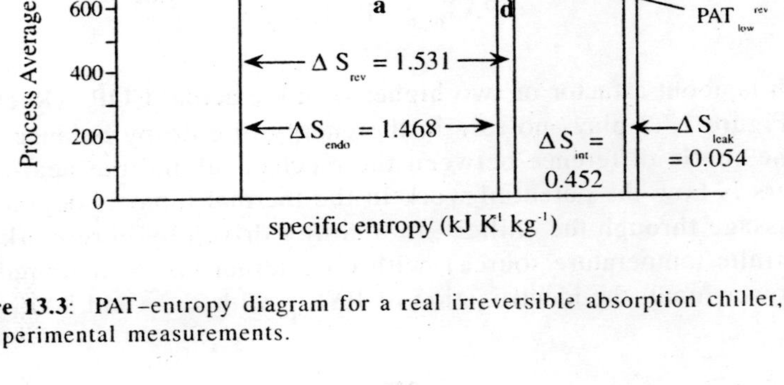

Figure 13.3: PAT–entropy diagram for a real irreversible absorption chiller, based on experimental measurements.

is as follows.

A) For the overall energy balance:

1) net heat rejection = $Q_{cond} + Q_{abs} + Q_{gen}^{leak} + Q_{evap}^{leak}$
= $PAT_{high}(\Delta S_{endo} + \Delta S_{int} + \Delta S_{leak})$
= sum of rectangles **a** + **b** + **c** + **g** + **h** + **i** =
= 2392 kJ kg^{-1}.

2) cooling capacity $Q_{evap} = PAT_{low}\ \Delta S_{endo}$ = rectangles **a** + **b** =
= 1099 kJ kg^{-1}

also $Q_{evap} = PAT_{low}^{rev}\ \Delta S_{rev}$ = rectangles **a** + **d** =
= 1099 kJ kg^{-1}.

3) thermal input Q_{gen} = net heat rejection – cooling capacity =
= 1304 kJ kg^{-1}.

4) $\mathrm{COP} = \dfrac{Q_{evap}}{Q_{gen}} = 0.84.$

B) The reversible limit:

1) minimum thermal input Q_{gen}^{min} = rectangles **b** + **c** + **e** + **f** =
= 216 kJ kg^{-1}.

2) reversible $\mathrm{COP} = \dfrac{Q_{evap}^{rev}}{Q_{gen}^{min}} = 5.08.$

C) Losses stemming from external and internal irreversibilities:

1) total external losses = rectangles **g** – (**d** + **e** + **f**) =
= 461 kJ kg^{-1}.

2) internal losses = rectangle **h** = 547 kJ kg^{-1}.

3) heat leak losses = rectangle **i** = 65 kJ kg^{-1}.

4) total losses = 1088 kJ kg^{-1}, of which 51% are solely internal losses.

5) $\mathrm{COP}_{endo} = \dfrac{Q_{evap}}{(PAT_{high} - PAT_{low})\Delta S_{rev}} = 1.62.$

which is about a factor of two higher than the actual COP. (Rectangle **j** in Figure 13.3 plays no role in the energy or entropy balance.)

One subtle difference between the mechanical and the heat-driven chillers is how the potential work in the thermal input is degraded in its passage through the chiller. In a chiller driven by pure work (*i.e.*, an infinite temperature source), with no external losses attributable to the power input, the identification of PAT_{high} with T_{cond}, and PAT_{low} with

Clearly, however, this cannot be a statement of general validity. In this chapter, we'll consider a few notable exceptions. Where appropriate, we would like to pose the associated modeling and analysis in simple thermodynamic terms as a challenge for future studies. The issue revolves around establishing how internal dissipation is a function of the principal operating variables in a physically-meaningful fashion.

We also note that the thermal modeling presented in earlier chapters should not be applied uncritically to cryogenic chillers. The domain of chiller operating variables can be far more expansive than for conventional mechanical and absorption chillers. This means that a substantial dependence of internal dissipation on the operating variables can arise. Even basic linear heat transfer may not be an adequate approximation at the low temperatures and over the large temperature differences encountered in cryogenic chillers. One challenge for future work is the development of the types of tools presented in this book to cryogenic refrigeration units.

In Sections B–D, we'll explore the consequences of instances of practical interest where the constant–ΔS_{int} approximation grows poor. We focus on three significantly different chiller types: thermoelectric, screw–compressor (mechanical), and multiple-stage absorption systems at part-load.

Sections E and F offer a brief review of two less conventional and unrelated chiller types that have been receiving increased attention in the commercial and research communities in recent years: adsorption chillers and vortex-tube chillers. In each case, we try to succinctly review the basic physics and engineering of their operation, and then proceed to try to understand their thermodynamic performance in terms of the modeling tools developed in earlier chapters.

B. THE THERMOELECTRIC CHILLER AS A CLEAR CUT CASE

Perhaps the simplest and most clear cut case of a chiller where internal losses vary strongly with cooling rate is the thermoelectric chiller. In Chapter 13, we cited the explicit dependence of entropy production rate and cooling rate on electrical current (Equations (13.11)–(13.13)). Even over the limited operating range of practical interest for thermoelectric chillers – which spans the regime from the point of maximum COP to the point of maximum cooling rate (a region that is clear from Figure 10.14) – the approximation of constant ΔS_{int} is a poor one. The problem also differs from that of conventional mechanical and absorption chillers in that there are no significant irreversibilities due to finite-rate heat exchange. The second irreversibility mechanism that dictates device performance is heat leak, with a dissipation rate that is constant at all cooling rates (for fixed reservoir temperatures and given material properties).

Simple modeling of the thermoelectric chiller is a solved problem, so no challenge is being raised here. Rather, we just note two points regarding this notable exception. First, the physical mechanism underlying internal dissipation (electrical resistance) has a markedly different functional dependence on cooling rate than the sources of internal loss in mechanical and absorption chillers.

And second, the thermoelectric device is not restricted to a limited range of relative cooling rates. Namely, once the thermoelectric chiller is built, we can access its entire theoretically-possible characteristic curve between zero and maximum cooling rate simply by varying the input current.

Contrast that with conventional commercial mechanical and absorption chillers. Manufacturers will typically design and construct them such that the maximum cooling rate at which they should be run roughly coincides with their maximum COP point. Due to mechanical limitations, there is a certain part-load cooling rate below which the machine should not or cannot be operated. Similarly, for mechanical chillers, although higher cooling rates are theoretically attainable, the chiller is not built to deliver them because the sacrifice in COP due to the thermal bottleneck at the heat exchangers would be excessive. The latter point was illustrated graphically in earlier chapters in the characteristic performance curve (1/COP plotted against 1/(cooling rate)) where beyond the point of maximum COP, COP decreases so rapidly with increasing cooling rate as to render this theoretically-possible high cooling rate region undesirable.

C. SCREW-COMPRESSOR CHILLERS

The evidence we've examined in Chapters 6–8 and 10–13 confirms that commercial reciprocating and centrifugal chillers appear to abide by the constant–ΔS_{int} approximation. The physical mechanism for internal losses in screw compressors is different than that in reciprocating and centrifugal units: it stems primarily from refrigerant leakages between the rotors (*i.e.*, refrigerant leaks back to the suction port via the clearances of the rotors and lobes), especially at high pressure ratios. As the thermal lift (*i.e.*, the difference between T_{cond} and T_{evap}) increases, the pressure ratio increases, and rotor leakages increase rapidly, with a concomitant drop in COP. Hence extending model predictions to screw-compressor devices must be checked carefully.

As a quantitative example, Table 14.1 summarizes experimental measurements from a nominal 1038 kW cooling rate commercial screw–compressor chiller [Mayakawa 1996]. Note that cooling rate and COP are measured as functions of refrigerant (rather than coolant) temperatures. This allows a test of the constant–ΔS_{int} approximation from the governing performance Equation (4.17), which can be rearranged as

Table 14.1: Experimental measurements from a nominal 1038 kW screw–compressor chiller, and the inferred/calculated values for $\Delta S_{int} + \Delta S_{leak}$.

Q_{evap} (kW)	1/COP	T_{evap} (K)	T_{cond} (K)	calculated $\Delta S_{int} + \Delta S_{leak}$ (kW K^{-1})
1656	0.071	283	293	0.20
1575	0.090	283	298	0.20
1502	0.111	283	303	0.20
1433	0.135	283	308	0.22
1372	0.163	283	313	0.25
1022	0.120	273	293	0.16
1000	0.135	273	298	0.15
953	0.163	273	303	0.17
915	0.196	273	308	0.20
874	0.235	273	313	0.25
674	0.163	263	293	0.11
657	0.200	263	298	0.15
635	0.220	263	303	0.14
602	0.277	263	308	0.21
562	0.339	263	313	0.27
434	0.233	253	293	0.11
418	0.277	253	298	0.14
402	0.331	253	303	0.18
389	0.394	253	308	0.22
366	0.458	253	313	0.26
157	0.558	233	293	0.16
147	0.680	233	298	0.20
136	0.831	233	303	0.24
125	1.018	233	308	0.28
111	1.275	233	313	0.33

$$\frac{Q_{\text{evap}}}{T_{\text{cond}}}\left[\frac{1}{\text{COP}}+1-\frac{T_{\text{cond}}}{T_{\text{evap}}}\right]=\Delta S_{\text{int}}+\Delta S_{\text{leak}}\,. \tag{14.1}$$

Table 14.1 includes the calculated value of the right-hand side of Equation (14.1) (*i.e.*, the measured value of the left-hand side of Equation (14.1)), which clearly is not approximately constant. The deviation from constancy increases as the inferred internal losses increase, *e.g.*, as COP decreases. (Since heat leaks are small relative to internal losses, the substantial dependence of entropy production here on operating conditions derives mainly from the internal dissipation.)

An alternative way of viewing the problem is the degree to which the single best-fit value of ΔS_{int} + ΔS_{leak} does not succeed in accounting for the measured COP values, which is illustrated in Figure 14.1.

The challenge for the screw-compressor chiller is basic modeling of internal losses such that the dependence of COP on cooling rate (and possibly other operating variables) is accounted for explicitly. Preferably, one would like to retain the simplicity of the analytic approach developed for reciprocating and centrifugal chillers in terms of formulae with which predictive, diagnostic and optimization studies can be performed with relative ease.

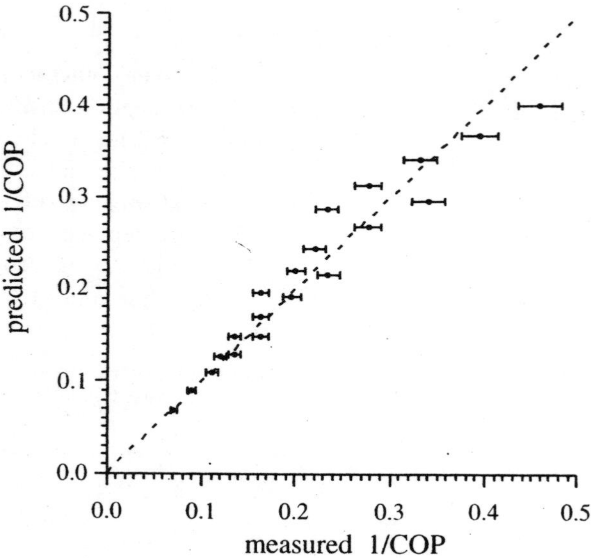

Figure 14.1: 1/COP predicted with Equation (14.1) based on a single best-fit value of the entropy production parameter, plotted against measured 1/COP for the screw–compressor chiller of Table 14.1. Experimental uncertainty bars are included.

D. REGENERATIVE ABSORPTION CHILLERS

The absorption machines we have analyzed cover a relatively narrow extent of generator temperatures. Hence we might expect these devices to exhibit a rate of internal entropy production that is relatively insensitive to part-load conditions.

This need not be the case for some regenerative multiple-stage absorption chillers where common part-load operating conditions can result in the generator covering a temperature range as broad as 60°C. To illustrate the trend, we list in Table 14.2 the simulated performance of a two-stage nominal 6.89 kW cooling rate ammonia–water absorption chiller with $T^{in}_{gen} = 190°C$ (and where, as in Chapters 9 and 12, the simulation was validated against experimental measurements before using it for performance predictions) [Tu 1997]. The rates of internal and external entropy production are listed at a few values of T^{in}_{gen}. They clearly show the non-constancy of ΔS_{int}. And unlike most chillers, this advanced design has also reduced internal losses to a level well below the external losses (but far from negligible).

Although the broad range of operation and the non-constancy of ΔS_{int} in Table 14.2 currently appear to be the exception rather than the rule, it raises the challenge of fundamental thermodynamic modeling such that these effects can be accounted for within a relatively simple analytic framework.

E. ADSORPTION CHILLERS

As a thermally-driven (as opposed to work-driven) machine, the adsorption chiller is qualitatively similar to the absorption chiller in the basic function of its components (see Figure 14.2 for a schematic). The absorption machine processes of (*a*) driving refrigerant out of solution with heat input at a generator, and of (*b*) absorbing refrigerant back into solution at a heat-rejecting absorber, are replaced by the adsorption device processes of (*i*) driving the refrigerant adsorbent out of a bed into the vapor phase at a heat-input desorber, and of (*ii*) adsorbing

Table 14.2: Rates of internal and external entropy production at 3 values of T^{in}_{gen} for a double-stage ammonia–water absorption chiller, based on simulated performance [Tu 1997].

T^{in}_{gen} (°C)	ΔS_{int} (W K^{-1})	ΔS_{ext} (W K^{-1})
190 (full load)	1.3	2.0
170 (part load)	0.9	1.8
145 (part load)	0.55	1.5

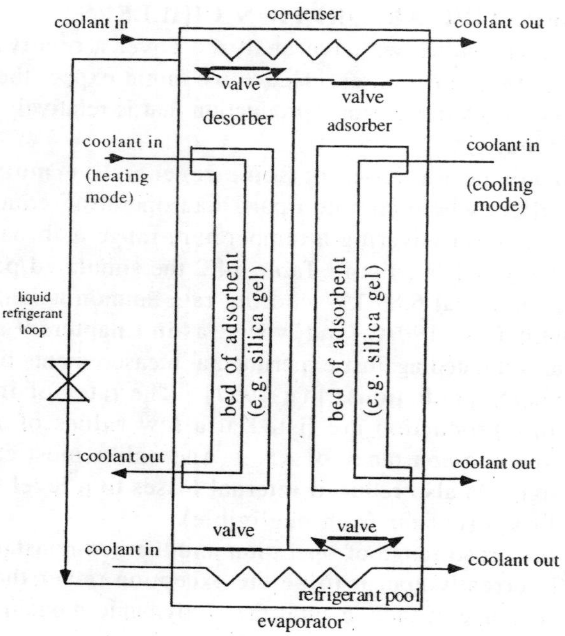

Figure 14.2: Schematic of an adsorption chiller. The cooling cycle is driven by heat input at the desorber where refrigerant is driven out of the adsorbent bed and heated. The condenser and evaporator play the same roles as in absorption and mechanical chillers. Refrigerant is adsorbed at a heat-rejecting adsorber. Due to the long times required for saturation in adsorption and desorption, batch processing is required. Valves at the entrance and exit of the adsorber and desorber restrict refrigerant flow as needed.

it back into the bed at a heat-rejecting adsorber. The condenser and evaporator serve the same functions as in absorption or mechanical chillers.

The COP of the adsorption cycle is defined as the ratio of cooling energy per cycle at the evaporator to the corresponding heat input per cycle at the desorber:

$$COP_{\text{adsorption chiller}} = \frac{Q_{\text{evap}}}{Q_{\text{desorber}}} \,. \tag{14.2}$$

Its reversible Carnot bound can be derived analogously to that of Equation (2.15) for the absorption chiller, the result being

$$\mathrm{COP}^{\mathrm{Carnot}}_{\text{adsorption chiller}} = \frac{\dfrac{1}{T_{\text{adsorber}}} - \dfrac{1}{T_{\text{desorber}}}}{\dfrac{1}{T_{\text{evap}}} - \dfrac{1}{T_{\text{adsorber}}}}. \tag{14.3}$$

Although the adsorption cycle is illustrated schematically in Figure 14.2 as a single-stage system, regenerative heat exchange and multi-bed designs can also be introduced to heighten chiller COP [Saha *et al* 1995].

A typical temperature–time trace for a single cycle in a two-bed water-silica gel adsorption chiller is graphed in Figure 14.3.

In the specific instance of adsorption chillers, the figure of merit that best reflects the energetic value of the chiller may not be the COP. The reason is that, unlike absorption chillers, adsorption chillers discard the unutilized heat from the input (desorber) stream. Often, this unexploited heat is at a sufficiently low temperature that it may not be economically worthwhile to recover it. But applications exist where utilization of the otherwise purged waste heat is worthwhile. In these cases, the COP would not indicate the nominal improvement in chiller performance linked to heat recovery, *e.g.*, by moving to a multi-bed configuration.

An alternative figure of merit in such cases could be a recovery efficiency, defined as the ratio of the cooling energy per cycle to the

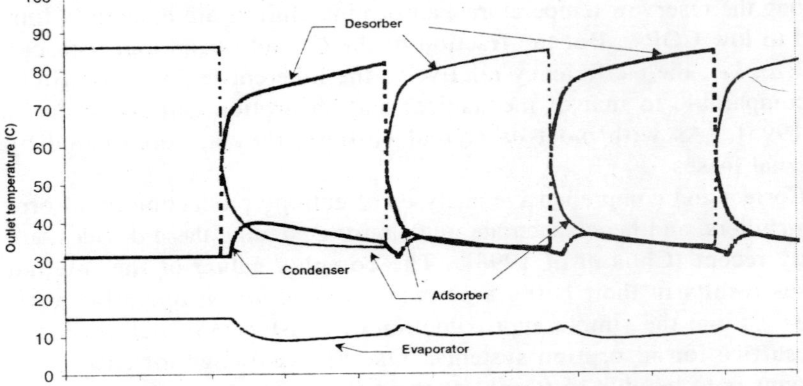

Figure 14.3: Temperature–time trace for a two-bed water-silica gel adsorption chiller, proceeding from turn-on to dynamic steady-state. The switching time between the two beds falls at 50 s into the cycle.

enthalpy of the heating stream relative to the environment [Chua *et al* 1998]. The recovery efficiency *does* indicate the advantages of multi-bed (regenerative) designs.

Some of the differences between the adsorption and absorption processes dictate basic distinctions between their respective chillers. First, the adsorption and desorption processes are relatively slow, requiring the order of minutes rather than seconds. Hence adsorption chillers must employ batch (as opposed to continuous) processing. Valves are introduced that can isolate the adsorber and desorber, as well as restrict the flow of refrigerant between each of them with the condenser and evaporator (as shown in Figure 14.2). Even the switching time between adsorption and desorption can be of the order of a minute. Therefore cycle times are very long compared to other refrigeration cycles.

Second, the internal irreversibilities and the finite-rate mass transfer losses linked to the adsorption and desorption processes are qualitatively and quantitatively different from those in absorption devices. Therefore, the manner in which adsorption chiller performance varies with the key operating variables is markedly different from the corresponding behavior of absorption chillers [Chua *et al* 1998].

Adsorption chillers are especially well-suited to heat utilization near environmental temperatures. The most promising adsorbent–desorbent pair identified to date for low-temperature heat input is silica gel and water. Until recently, the practical input temperature range was 60–80°C. With the latest advances in adsorption chiller technology, that range has now been decreased to 30–50°C [Saha *et al* 1995].

Because of the low heat input temperatures and the small differences among the reservoir temperatures, adsorption chillers are inherently limited to low COPs. But the fraction of the Carnot bound that they can realize, *i.e.*, their efficiency relative to the inherent fundamental limit, is comparable to that of mechanical and absorption chillers [Saha *et al* 1995]. As with most other real chillers, they are dominated by internal losses.

Correct and comprehensive analyses of entropy production in adsorption chillers, and hence accurate models for designing these devices, are fairly recent [Chua *et al* 1998]. The complex nature of the internal losses results in their being a perceptible function of operating variables, so that the simple approximation of constant ΔS_{int} will probably not suffice for adsorption systems. One key challenge for adsorption systems is to be able to formulate their governing equations in a form that allows the type of predictive, diagnostic and optimization tools developed in this book for other chiller types.

F. VORTEX-TUBE CHILLERS

F1. Device description and how vortex motion creates a cooling effect

An intriguing and somewhat different cooling device that has found a niche market for small-scale spot-cooling applications is the vortex-tube chiller. Vortex motion, *i.e.*, a fluid rotating about an axis, appears in a host of common phenomena such as tornados, motion in a stirred tea cup, eddies at the back of a fast cyclist, and many more. If the vortex is confined in a rigid cylindrical tube, then the device produces a marked temperature splitting wherein one end of the tube cools down, and the other warms up, relative to ambient conditions.

Vortex-tube chillers use an ordinary compressed air supply as a power source, and contain no moving parts. They create one hot and one cold stream of air. Operation is illustrated schematically in Figure 14.4. Compressed air, typically in the range of 1.4–7 bar, is introduced tangentially into a cylindrical tube of large aspect ratio. With speeds up to 10^6 revolutions per minute, the air stream twists in a hot outer loop around the periphery (akin to a whirlpool).

At the far end of the tube, a controllable fraction is bled as heated air through a needle valve. The remaining air is forced back through the center of the incoming air stream (*i.e.*, through the outer vortex) in an inner stream at a slower speed. This inner stream dissipates kinetic energy as heat to the outer stream, and exits the vortex tube as cooled air at the opposite end. The hot and cold streams comprise a primary

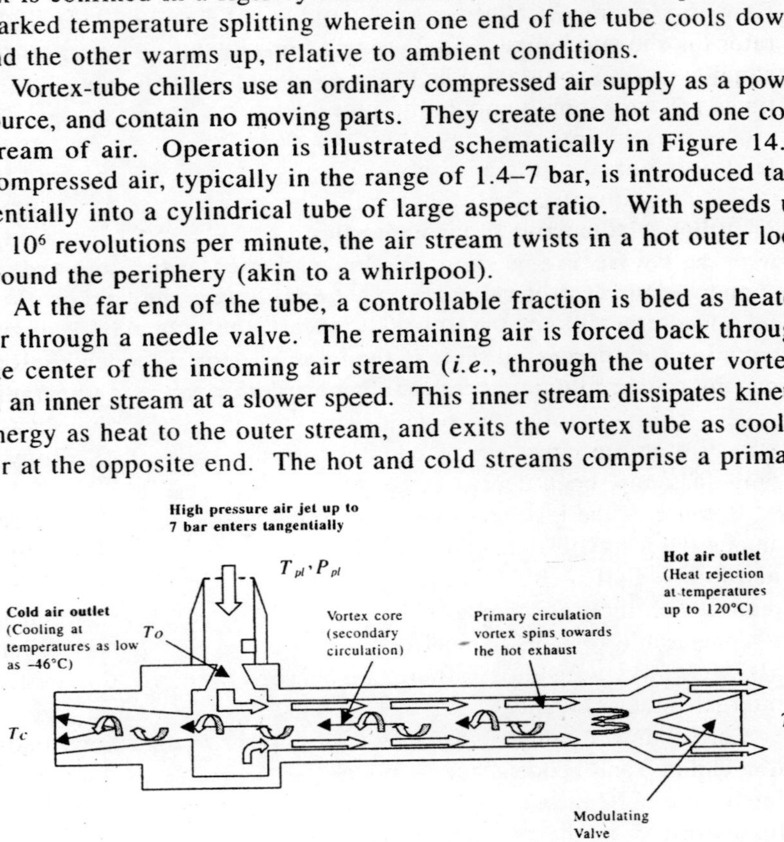

Figure 14.4: Schematic of a vortex-tube chiller. Compressed air at pressure p_{pl} and temperature T_{pl} is introduced to the vortex tube tangentially through the plenum at mass flow rate m_o and cools by expansion to temperature T_o. The outer air stream spirals toward the hot end, heating up along the way, where a prescribed fraction is extracted via a control valve. The remaining rotating air is forced back as an inner stream through the center of the peripheral circulation. The inner stream cools via dissipation and heat transfer to the outer flow and exits the tube as cold air.

loop. Embedded within that is a secondary flow loop.

The minimum pressure of 1.4 bar is around the threshold for establishing adequately high-speed flow within the tube. Above the maximum recommended pressure of 7 bar, one encounters the onset of detrimental shock wave formation.

The inner stream in vortex flow is commonly observed in a stirred tea cup as tea leaves ascend along the axis and descend near the wall. This derives from the retardation of rotating fluid motion by friction at the wall; hence centrifugal acceleration is greater at the top of the cup than at the bottom. With a compressible working fluid such as air, rapid motion results in the compression of the air due to the centrifugal forces near the periphery, and in the expansion of the air near the tube's axis. The secondary loop moves between regions of high and low pressure.

F2. Chiller performance characteristics

Decreasing the fraction of the total flow extracted at the cold end, y, (*i.e.*, opening the control valve at the hot end) lowers the cold air flow but also decreases its temperature. In concert, the temperature of the delivered hot air decreases. This is the basic tradeoff between cooling power and refrigeration temperature. Sample performance is illustrated

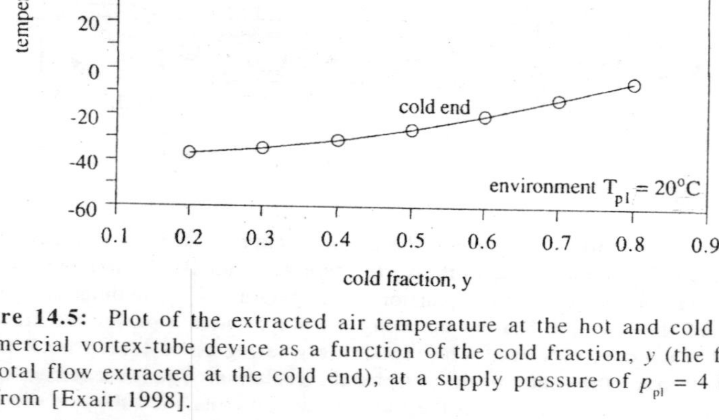

Figure 14.5: Plot of the extracted air temperature at the hot and cold ends of a commercial vortex-tube device as a function of the cold fraction, y (the fraction of the total flow extracted at the cold end), at a supply pressure of p_{pl} = 4 bar. Data are from [Exair 1998].

in Figure 14.5 for a commercial device. Typical vortex tube dimensions are a length of around 150 mm and a diameter of about 15 mm.

Although the vortex tube can equally well be used as a heat pump by extracting the useful effect at the hot end, its commercial value has been realized for spot-cooling applications, *e.g.*, the cooling of electronic controls, soldered parts, electronic components, heat seals, etc. Current commercial units driven with compressed air at room temperature can generate cold-air delivery temperatures as low as –50°C (and hot-air temperatures as high as 130°C). Obtainable cooling rates are in the range 40–3000 W. Aside from no moving parts, vortex-tube chillers offer the advantages of requiring no electricity, being small and lightweight, demanding little or no maintenance, a broad range of easily adjustable temperatures, durability (usually being made of stainless steel), and rapid start-up.

F3. Modeling the vortex-tube chiller

Although the fluid dynamics of the vortex tube have been studied, the thermal aspects of its behavior were analyzed only recently [Ahlborn *et al* 1998]. The creation of a thermal lift (temperature splitting) in the vortex tube can be viewed in terms of conventional chiller cycles. In this case, the cycle is created by the secondary circulation in the forced vortex motion, which:

(1) absorbs heat near the tube's axis at low pressure,

(2) is compressed,

(3) rejects heat in the periphery to the entering air that has cooled upon expansion, and

(4) expands prior to providing the cooling effect to the primary loop.

All 4 stages are comparable to those of conventional mechanical chiller cycles (as depicted in Chapter 2). And as noted earlier, a considerable fraction of the kinetic energy of the gas is dissipated into heat due to wall friction in the hot stream.

The known input variables are the temperature and pressure in the plenum (T_{pl} and p_{pl}, respectively), the cold fraction y, and the total mass flow rate m_o. The variables that need to be calculated with the model of [Ahlborn *et al* 1998] are the hot-end and cold-end delivery temperatures (T_h and T_c, respectively), and the gas temperature immediately after expansion in the plenum, T_o. The model predictions of [Ahlborn *et al* 1998] for T_h and T_c were validated against experimental measurements. Accordingly, we proceed with that model as a realistic tool for predicting T_h, T_c and T_o in the analysis that follows.

F4. The *external* perspective of the chiller

One can view the chiller from an external or an internal perspective. The external perspective is the one of practical interest to the consumer and manufacturer. The reference temperature for heat transfer is T_{pl}, so the cooling power Q_{cold} is

$$Q_{\text{cold}} = m_o C y (T_{pl} - T_c) \tag{14.4}$$

where C is the gas specific heat. The work input is that required to compress air from ambient conditions to those of the plenum: pressure p_{pl} and temperature T_{pl}. (Consumers will typically be impervious to the work input *per se*, and relate to the price of compressed air.)

For typical realistic values of these variables [Exair 1998, Ahlborn *et al* 1998], we estimate the COP of the chiller from the external perspective to be no greater than around 0.1. Such an inefficient chiller cannot compete with conventional chillers for standard cooling applications, as reflected by commercial realities. The extreme inefficiency can be understood from the poor coupling of the work available in the compressed air into the work induced in the vortex's loops. Detailed modeling of the energy flows and entropy balances is not a simple exercise and remains to be developed.

F5. The *internal* perspective of the chiller

The external perspective may be the one of interest to the manufacturer, consumer and cooling engineer. But the *internal* perspective can be of interest for a basic understanding of chiller dynamics. It is also from the internal perspective that we can view the vortex-tube chiller in the same manner in which we analyzed conventional chillers in Chapters 4–6. The secondary flow is now the refrigerant motion of a *cyclic* chiller. It experiences the 4 stages noted above (heat absorption, compression, heat rejection and expansion), in analogy to conventional chiller cycles. The primary loop plays the role of the coolant. One key difference relative to ordinary chillers is that the vortex coolant constitutes a single loop.

From the *internal* perspective, the energy balances are not referenced to T_{pl}. Rather, they should be calculated based upon the temperatures entering and leaving the heat rejection and heat absorption regions in the vortex tube, while accounting for the different flow rates in those regions. While a comprehensive analysis also remains a topic for future investigation, we *can* estimate the COP of the internal chiller. We use a prime (′) to denote energy flows from the internal perspective, in order to distinguish them from the external perspective.

The heat rejection from the internal chiller Q'_{hot} is

$$Q'_{\text{hot}} = m_o C (T_h - T_o) \tag{14.5}$$

and the heat absorption from the primary loop Q'_{cold} is

$$Q'_{\text{cold}} = m_o C\, y(T_h - T_c) . \tag{14.6}$$

From the First Law, the rate of work input W' to the secondary flow from the primary flow (not to be confused with the work input from the external perspective of compressing ambient air to plenum conditions) is

$$W' = Q'_{\text{hot}} - Q'_{\text{cold}} \tag{14.7}$$

and

$$\text{COP}' = \frac{Q'_{\text{cold}}}{W'} . \tag{14.8}$$

Using flow rates and temperatures characteristic of real vortex chillers, we find that COP′ is an order of magnitude greater than the COP from the external perspective. To wit, COP′ is in the range 0.1–1.0 and occasionally slightly higher. From the internal perspective, the losses associated with coupling the work in the inlet flow into the secondary loop are filtered out. Hence COP′ is dramatically larger than the COP from the external perspective.

This is somewhat analogous to open-type reciprocating compressors, where the compressor is driven by an external belt-and-pulley assembly. The mechanical dissipation due to slippage in the external driving machinery can readily be distinguished from the internal dissipation and heat-exchange losses of the reciprocating chiller proper. Hence for open-type reciprocating compressors, one can similarly adopt an external-*vs*-internal perspective in chiller analysis.

F6. Characteristic chiller plots and their interpretation

The internal perspective of the vortex chiller can be analyzed simply with the tools developed in earlier chapters. Specifically, a characteristic chiller plot should reveal basic information about the dominant irreversibilities. First, however, we can anticipate the principal features from the simple physical picture of the vortex chiller, and then check them against the actual data. Specifically:

a) The long spiraling path of a fluid element over the course of one chiller cycle provides ample opportunity for good heat transfer between the primary and secondary loops. So the irreversibilities that stem from finite-rate heat exchange should be dwarfed by the considerable losses due to fluid friction. That would also mean that refrigerant and coolant temperatures can be approximated as equal.

b) The rate of internal dissipation should depend primarily on the pressure drop across the tube, and not on cooling rate. Hence to a good approximation, ΔS_{int} should be constant for a given value of p_{pl}, but should change roughly in proportion to $\ln(p_{pl})$ (*i.e.*, the entropy change associated with the pressure drop of an ideal gas).

c) Because internal losses should be far greater than external losses, the chiller plots should be a series of straight lines, with slopes approximately proportional to $\ln(p_{pl})$. The extrapolation of those lines toward the high cooling rate regime should pass through the origin, precisely because external irreversibilities should be negligible (*relative* to fluid friction losses).

The cooling rate of the vortex tube is varied by changing the cold fraction y. But varying y also changes T_h and T_c. Therefore, the proper plot should be 1/COP against T_h/Q'_{cold} (rather than against $1/Q'_{\text{cold}}$) toward checking our predictions.

Using Equations (14.5)–(14.8), we plot 1/COP′against a non-dimensionalized abscissa, $\dfrac{T_h}{y(T_h - T_c)}$, (the latter being $m_o C T_h / Q'_{\text{cold}}$) in Figure 14.6. These plots are based on *measured*, rather than calculated, points [Exair 1998]. All the expected features noted above for chiller behavior are borne out.

For the range of cooling rates to which vortex-tube chillers appear to be limited, the heat-exchange irreversibility appears to be small compared to that of fluid friction. Since the high-cooling-rate limit is dictated by the avoidance of shock-wave formation, it appears likely that these curves will remain restricted to the linear, *relatively* low cooling rate side of the characteristic chiller performance curve.

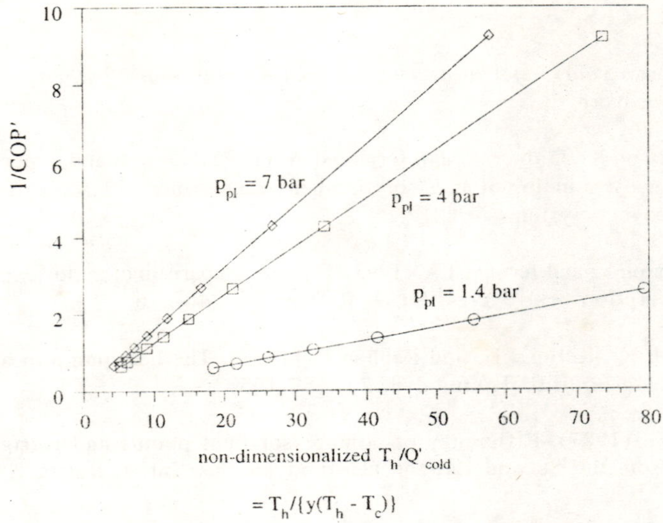

Figure 14.6: Characteristic chiller plot for the *internal* perspective of the vortex-tube chiller. The plotted points are based on experimental measurements [Exair 1998]. The linearity of the curves indicates the dominant role of internal dissipation from fluid friction, and the *relatively* small role of finite-rate heat-exchange losses. Also, the slope of the lines, which should correspond to the rate of internal entropy production, increases with the pressure of the compressed air (and hence with the pressure drop across the tube), in accordance with the physical picture.

REFERENCES

"If you wish your merit to be known, acknowledge that of other people." - Chinese proverb

Abrahamsson K., Gidner A. and Jernqvist Å. (1995). Design and experimental performance evaluation of an absorption heat transformer with self-circulation. Heat Recovery Systems & CHP 15, 257-272.

Abrahamsson K. and Jernqvist Å. (1993). Carnot comparison of multi-temperature level absorption heat cycles. Int. J. Refrig. 16, 240-246.

Ahlborn B.K., Keller J.U. and Rebhan E. (1998). The heat pump in a vortex tube. J. Non-Equilib. Thermodyn. 23, 159-165.

Alefeld G. (1987). Efficiency of compressor heat pumps and refrigerators derived from the Second Law of Thermodynamics. Int. J. Refrig. 10, 331-341.

Andresen B., Salamon P. and Berry R.S. (1977). Thermodynamics in finite time: extremals for imperfect heat engines. J. Chem. Phys. 66, 1571-1577.

ARI Air-conditioning and Refrigeration Institute (1985). Standard 325: Standard for ground water-service heat pumps. ARI, Arlington, VA.

ARI Air-conditioning and Refrigeration Institute (1986a). Standard 550: Centrifugal or Rotary Screw Water-Chilling Packages. ARI, Arlington, VA.

ARI Air-conditioning and Refrigeration Institute (1986b). Standard 560: Absorption Water-Chilling Packages. ARI, Arlington, VA.

ARI Air-conditioning and Refrigeration Institute (1986c). Standard 590: Standard for reciprocating water-chilling packages. ARI, Arlington, VA.

ARI Air-conditioning and Refrigeration Institute (1986d). Standard 320: Standard for water-source heat pumps. ARI, Arlington, VA.

ARI Air-conditioning and Refrigeration Institute (1987). Standard 470: Desuperheater water heaters. ARI, Arlington, VA.

ASHRAE (The American Society of Heating, Refrigerating and Air-Conditioning Engineers) (1975). Standard 413-75. Standard measurement guide: engineering analysis of experimental data. ASHRAE Inc., Washington, D.C.

ASHRAE (1986). Thermodynamic Properties of Refrigerants. ASHRAE Inc., Washington, D.C.

ASHRAE (1997). GPC-14P. A publication guide to measure energy savings from energy conservation retrofits. ASHRAE Inc., Atlanta, GA.

ASHRAE (1998). ASHRAE Handbook, Fundamentals Volume (SI Units). ASHRAE Inc., Atlanta, GA.

Austin S.B. (1991). Optimal chiller loading. ASHRAE J. 33, 40-43.

Bartana A., Kosloff R. and Tannor D.J. (1993). Laser cooling of molecular internal degrees of freedom by a series of shaped pulses. J. Chem. Phys. 99, 196-210.

Beyenem A., Güven H., Jaedat Z. and Lowrey P. (1994). Conventional chiller performance simulation and field data. Int. J. Energy Res. 18, 391-399.

Bong T.Y., Ng K.C. and Lau K.O. (1989). Water-to-water heat pump test facility. Heat Recovery Systems & CHP 9, 133-141.

Bong T.Y., Ng K.C. and Lau K.O. (1990). Test facility for water-cooled water chiller. ASHRAE Trans. 96, 205-212.

Brandemuehl M.J. (1995). Methodology development to measure in-situ chiller, fan and pump performance. Systems Energy Utilization. ASHRAE Research Project RP-827, TC 9.6. ASHRAE Inc., Atlanta, GA.

Carrier Air Conditioning Co. (1962). Operation and maintenance of an absorption chiller for Central Baptist Hospital, Kentucky. Customer's order No. EM5001, Carrier job No. 246-0E6-018. Carrier Corp., USA.

Carrier Corp. (1971). Operation and maintenance instructions - hermetic centrifugal liquid chillers. Model 19DG.

Carrier International Corp. (1984). Packaged cooling units (small air-cooled chillers). Models 50DP016-020, Catalog No. 005-013. Syracuse, NY.

Çengel Y.A. and Boles M.A. (1989). Thermodynamics: An Engineering Approach. McGraw-Hill, NY.

Chua H.T. (1995). Performance analysis of vapour compression chillers. M.Eng. thesis, Department of Mechanical & Production Engineering, National University of Singapore.

Chua, H.T. (1999). Universal thermodynamic modelling of chillers: special application to adsorption chillers. Ph.D. thesis, Department of Mechanical & Production Engineering, National University of Singapore.

Chua H.T., Gordon J.M., Ng K.C. and Han Q. (1997). Entropy production analysis and experimental confirmation of absorption systems. Int. J. Refrig. 20, 179-190 (1997).

Chua H.T., Ng K.C. and Gordon J.M. (1996). Experimental study of the fundamental properties of reciprocating chillers and its relation to thermodynamic modeling and chiller design. Int. J. Heat Mass Transfer 39, 2195-2204.

Chua H.T., Ng K.C., Malek A., Kashiwagi T., Akisawa A. and Saha B.B. (1998). Entropy analysis of two-bed silica gel-water, non-regenerative adsorption chillers. Journal of Physics D: Applied Physics 31, 1471-1477.

Chuang C.C. and Ishida M. (1990). Comparison of three types of absorption heat pumps based on energy utilization diagrams. ASHRAE Trans. Part 2, 275-281.

Ebara Corp. (1995). Operation and Maintenance Manual of Single- and Two-stage Absorption Chillers. 2-1 Honfujisawa, 4-chome, Fujisawa-shi 251, Japan.
Exair Corp. (1998). Exair Vortex Tubes. 1250 Century Circle N., Cincinnati, OH, USA.

Garrett S.L. and Hofler T.J. (1992). Thermoacoustic refrigeration. ASHRAE J. 34, 28-36.

Geva E. and Kosloff R. (1996). The quantum heat engine and heat pump: an irreversible thermodynamic analysis of the three-level amplifier. J. Chem. Phys. 104, 7681-7698.

Goldsmid H.J. (1960). Applications of Thermoelectricity. Methuen, NY.

Gordon J.M. and Huleihil H. (1992). General performance characteristics of real heat engines. J. Appl. Phys. 72, 829-837.

Gordon J.M. and Ng K.C. (1994a). Thermodynamic modeling of reciprocating chillers. J. Appl. Phys. 75, 2769-2774.

Gordon J.M. and Ng K.C. (1994b). A general thermodynamic model for absorption chillers: theory and experiment. Heat Recovery Systems & CHP 15, 73-83.

Gordon J.M. and Ng K.C. (1995). Predictive and diagnostic aspects of a universal thermodynamic model for chillers. Int. J. Heat Mass Transfer 38, 807-818.

Gordon J.M., Ng K.C. and Chua H.T. (1995). Centrifugal chillers: thermodynamic modeling and a diagnostic case study. Int. J. Refrig. 18, 253-257.

Gordon J.M., Ng K.C. and Chua H.T. (1997). Optimizing chiller operation based on finite-time thermodynamics: universal modeling and experimental confirmation. Int. J. Refrig. 20, 191-200.

Gordon J.M., Ng K.C. and Chua H.T. (1999). Simple thermodynamic diagrams for real refrigeration systems. J. Appl. Phys. 85, 641-646.

Gordon J.M., Ng K.C., Chua H.T. and Lim C.K. (2000). How varying condenser coolant flow rate affects chiller performance: thermodynamic modeling and experimental confirmation. Applied Thermal Engineering, 20, 1149-1159.

Grazzini G. (1993). Irreversible refrigerators with isothermal heat exchanges. Int. J. Refrig. 16, 101-106.

Herold K.E., Radermacher R. and Klein S.A. (1996). Absorption Chillers and Heat Pumps. CRC Press, Boca Raton.

Holman, J.P. (1992). Heat Transfer. 7th edition. McGraw-Hill, Singapore.

Ioffe A.F. (1957). Semiconductor Thermoelements and Thermoelectric Cooling. Infosearch, London.

Jernqvist Å., Abrahamsson K. and Aly G. (1992a). On the efficiencies of absorption heat transformers. Heat Recovery Systems & CHP 12, 323-334.

Jernqvist Å., Abrahamsson K. and Aly G. (1992b). On the efficiencies of absorption heat pumps. Heat Recovery Systems & CHP 12, 469-480.

Kreider J.F. and Rabl A. (1994). Heating and Cooling of Buildings: Design for Efficiency. McGraw-Hill, NY. Ch. 10, Sect. 10.5, pp. 497-500.

Leverenz D.J. and Bergan N.E. (1983). Development and validation of a reciprocating chiller model for hourly energy analysis programs. ASHRAE Trans. 89(1A), 156-174.

Liang H. and Kuehn T.H. (1991). Irreversibility analysis of a water-to-water mechanical-compression heat pump. Energy Int. J. 16, 883-896.

Liu K., Güven H., Beyenne A. and Lowrey P. (1994). A comparison of the field performance of thermal energy storage (TES) and conventional chiller systems. Energy Int. J. 19, 889-900.

Mayakawa Manufacturing Co. Ltd. (1996). Catalog No. 010S098Y 1E-SIC, Model SRM-200. 2-13-1, Botan, Koto-Ku, Tokyo, Japan.

Mayhew Y.R. and Rogers G.F.C. (1971). Thermodynamic and Transport Properties of Fluids (SI Units). Basil Blackwell, Oxford.

Mills A.F. (1992). Heat Transfer. Irwin Press, Homewood, IL, USA.

Ng K.C., Chua H.T., Bong T.Y., Lee T.Y., Lee S.S. and Lee T.K. (1994). Experimental and theoretical analysis of water-cooled chiller. I.E.S. J. Singapore 34, 45-54.

Ng K.C., Chua H.T. and Han Q. (1997a). On the modeling of absorption chillers with external and internal irreversibilities. Appl. Thermal Eng. 17, 413-425.

Ng K.C., Chua H.T., Ong W., Lee S.S. and Gordon J.M. (1997b). Diagnostics and optimization of reciprocating chillers: theory and experiment. Appl. Thermal Eng. 17, 263-276, + Erratum 17, 601-602.

Ng K.C., Chua H.T. and Ong A.S. (1997c). Experimental verification of a diagnostic model for reciprocating chillers. Proc. Inst. Mech. Eng. 211 E, 259-265.

Ng K.C., Tu K., Chua H.T., Gordon J.M., Kashiwagi T., Akisawa A. and Saha B.B. (1998a). The role of internal dissipation and process average temperature in chiller performance and diagnostics. J. Appl. Phys. 83, 1831-1836.

Ng K.C., Tu K., Chua H.T., Gordon J.M., Kashiwagi T., Akisawa A. and Saha B.B. (1998b). Thermodynamic analysis of absorption chillers: internal dissipation and process average temperature. Appl. Thermal Eng. 18, 671-682.

Ng, K.C., Chua, H.T., Han, Q., Kashiwagi, T., Akisawa, A. and Tsurusawa, T. (1999). Thermodynamic modeling of absorption chiller and comparison with experiments. Heat Transfer Eng. 20 (2), 41-52.

Saha B.B., Boelman E.C. and Kashiwagi T. (1995). Computational analysis of an advanced adsorption-refrigeration cycle. Energy 20, 983-994.

Stoecker W.F. and Jones J.W. (1982). Refrigeration and Air Conditioning. 2nd edition. McGraw-Hill, Singapore.

Summerer F. (1996). Evaluation of absorption cycles with respect to COP and economics. Int. J. Refrig. 19, 19-24.

Swift G.W. (1988). Thermoacoustic engines. J. Acoust. Soc. Am. 84, 1145-1180.

Taniguchi F. *et al* (1996). The development of ammonia-water absorption chiller with GAX. Proc. of the 30th Japanese Joint Conf. on Refrigeration and Air-conditioning. Paper No. 40. Tokyo, Japan. pp. 157-160.

Toyo Carrier Engineering Co. (1989). 38PE 40HQ AQ TQ split system cooling units, 50 Hz, cooling 18.6-130.2 kW. Publication ECR9105-1(S). Tokyo, Japan.

Toyo Carrier Engineering Co. (1991). 30 HKA HK HR packaged hermetic reciprocating chillers, 50 Hz, 45.4 to 461 kW, 15 to 160 tons. Publication EPD9107-1(S). Tokyo, Japan.

Trane Co. (1989). Single stage absorption cold generator: 101 to 1660 tons. Catalog ABS-DS-1. La Crosse, WI.

Trane Co. (1990). Cold generator reciprocating liquid chillers: 70 to 120 tons water-cooled and condenserless. Catalog CG-DS-4, Publication PL-RF-CG-000-DS-4-690. La Crosse, WI.

Trane Co. (1992). Air cooled reciprocating liquid chillers, series CGAV 330 kW through 1180 kW. Societé Trane Publication C47SD603E-0892. Golbey, France.

Trane Co. (1996). Refrigeration data catalog for CenTraVac liquid (water-cooled) chiller, CTV-DS-15-296. LaCrosse, WI, USA.

Tu K. (1997). Waste-heat powered absorption chillers: theoretical modelling. M.Eng. thesis, Department of Mechanical & Production Engineering, National University of Singapore.

Wetzel M. and Herman C. (1997). Design optimization of thermoacoustic refrigerators. Int. J. Refrig. 20, 3-21.

Yazaki Resources Co. (1979). Installation and Service Manual, WFC-600 Model. Shizuoka-ken, Japan.

Ziegler B. and Trepp Ch. (1984). Equation of state for ammonia-water mixtures. Int. J. Refrig. 7, 101-106.

Zhou C.Z. and Machielsen C.H.M. (1996). Performance of high-temperature absorption heat transformers using alkitrate as the working pair. Appl. Thermal Eng. 16, 255-262.

Index

A

B

C

D

E

F

G

H

I

L

M

N

P

Q

R

S

T

U

V

W